Evaporation in the Uplands

Evaporation in the Uplands

Ian R. Calder
Institute of Hydrology
Wallingford
UK

JOHN WILEY & SONS
Chichester · New York · Brisbane · Toronto · Singapore

Other Wiley Editorial Offices

John Wiley & Sons, Inc., 605 Third Avenue,
New York, NY 10158-0012, USA

Jacaranda Wiley Ltd, G.P.O. Box 859, Brisbane,
Queensland 4001, Australia

John Wiley & Sons (Canada) Ltd, 22 Worcester Road,
Rexdale, Ontario M9W 1L1, Canada

John Wiley & Sons (SEA) Pte Ltd, 37 Jalan Pemimpin   05-04,
Block B, Union Industrial Building, Singapore 2057

Library of Congress Cataloging-in-Publication Data:

Calder, Ian R.
    Evaporation in the uplands / Ian R. Calder.
      p.   cm.
    Includes bibliographical references.
    ISBN 0 471 92487 3
    1. Evaporation (Meteorology)—Great Britain.   2. Rain and
rainfall—Great Britain—Measurement.   I. Title.
QC915.7.G7C35   1990   89–16647
551.57′2′0941—dc20   CIP

British Library Cataloguing in Publication Data:

Calder, Ian R.
    Evaporation in the uplands.
    1. Afforestation. Effects on water resources
    I. Title
    333.91

    ISBN 0 471 92487 3

Typeset by Photo·graphics, Honiton, Devon
Printed in Great Britain by Courier International Ltd., Tiptree, Essex

Contents

Preface   ix
Acknowledgements   xi
Symbols   xiii

CHAPTER ONE   Introduction   1
  1.1 Frank Law and the Stocks Reservoir controversy   1
  1.2 Frank Law's studies   2
  1.3 The Plynlimon catchment results   5
  1.4 The Plynlimon forest lysimeter   7
  1.5 The Thetford micrometeorological and soil
        physical studies   11
  1.6 The continuing controversy   13

CHAPTER TWO   Physics and Physiology of
                  Evaporation   16
  2.1 Introduction   16
  2.2 The Dalton equation   17
  2.3 The energy balance equation   19
  2.4 The 'del' approximation and the combination
        equation   19
  2.5 The Penman equation   21
  2.6 Physiological resistance—the Penman–Monteith
        equation   25
  2.7 Surface temperature   26
  2.8 Evaporation of intercepted precipitation   28
  2.9 Transport theory   29
  2.10 Evaporation–atmosphere interaction   32

CHAPTER THREE   The Interception Process   40
  3.1 Introduction   40
  3.2 The Rutter interception model   40
  3.3 The stochastic interception model   41
  3.4 The simplified 'daily' stochastic model   47

vi   Contents

CHAPTER FOUR   Further Transpiration Studies in
                    the Uplands   49
4.1 Introduction   49
4.2 The Stocks heather and grass lysimeters   49
4.3 Grass drainage lysimeters—Plynlimon   54
4.4 Soil moisture studies of transpiration   56
4.5 The Sneaton Moor heather studies   58

CHAPTER FIVE   Rainfall Interception from Short
                    Vegetation—The 'Wet Surface'
                    Lysimeter Experiment   60
5.1 Experimental design   60
5.2 Theory   61
5.3 Sites   61
5.4 Aerodynamic resistance results   67
5.5 Canopy drainage measurements   69
5.6 Evaporation in partially wet conditions   72
5.7 Wetting characteristics of vegetation   72

CHAPTER SIX   Gamma-Ray Attenuation Studies of
                    Rainfall Interception from Forests   76
6.1 Introduction   76
6.2 Experimental design   76
6.3 Experimental method   78
6.4 Description of sites   81
6.5 Canopy density profiles   82
6.6 Wet canopy observations   83
6.7 Canopy drainage function   84
6.8 Evaporation from partially wet canopies   86
6.9 Dependence of drainage parameters and
    aerodynamic resistance on wind speed   87

CHAPTER SEVEN   Plastic-Sheet Net-Rainfall Gauge
                    Measurements of Forest
                    Interception   89
7.1 Introduction   89
7.2 Studies in Wales and Scotland   92
7.3 Tree-thinning experiment—Plynlimon   97

CHAPTER EIGHT   Evaporation from Snow-Covered
                    Vegetation   100
8.1 Evaporation from a 'smooth' snow pack   100
8.2 Snow interception from forests—introduction   101
8.3 Snow interception from forests—experimental
    results   102
8.4 The Aviemore snow event of 11–13 December
    1983   105
8.5 Snow interception from forests—modelling   107
8.6 Model predictions   112

CHAPTER NINE   Predicting the Effects of Land
                  Use Change—Evaporation Models   114
  9.1 Introduction   114
  9.2 Research models   115
  9.3 Simple annual evaporation model   116
  9.4 Development of the daily interception model 118
  9.5 Seasonal evaporation model for forest, heather
      and grass   119

CHAPTER TEN   Implications of Upland
                  Afforestation—A Case Study: The
                  Crinan Canal   124
  10.1 Introduction   124
  10.2 The Crinan Canal reservoirs   124
  10.3 Evaporation measurements   126
  10.4 Modelling studies   126

CHAPTER ELEVEN   Remaining Questions   131
  11.1 Introduction   131
  11.2 Precipitation measurements   132
  11.3 Transport theory   132
  11.4 Scale of the evaporation–atmosphere interaction 134
  11.5 Variation of evaporation with altitude   134
  11.6 Summary   137

References   138

Appendix   145

Index   147

Preface

Since the mid 1950s, when Frank Law published the results of his studies from Stocks Reservoir, a controversy has raged concerning the effects of upland afforestation on evaporation and water resources of the UK.

Originally the concern centred on the effects of spruce afforestation of upland moorland water catchments which were being used for supply purposes; then, by the late 1970s, the deleterious effects of afforestation on hydroelectric power generation were recognised. More recently, with food surpluses being generated within the European Community and the possibility of reductions in subsidies to farmers and the consequent probability of marginal agricultural lands (not always in the uplands) being afforested, the hydrological effects caused by extensive broad leaf plantations are becoming an important issue.

Fortunately, although the hydrological effects of afforestation remain a contentious issue worldwide, there is now a body of evidence to provide information for resource evaluation in temperate climates; in particular for the uplands of the UK.

By the late 1970s, studies carried out by the Institute of Hydrology and other organizations had validated Law's conclusion concerning enhanced evaporation rates from upland forests. Doubts were expressed, however, that because most of the experimental studies had been carried out in the relatively warm climates of Wales and England, primarily on grass and forest vegetation, results could not necessarily be applied in Scotland, where the climate is cooler and snow can form a significant component of the annual precipitation. Moreover, heather rather than grass moorland predominates. Further studies, involving both process and catchment experiments were executed to quantify the effects. The research was funded by a Consortium which included the British Waterways Board, Department of Energy, North of Scotland Hydro-Electric Board, Scottish Development Department, Forestry Commission, Department of the Environment and the Water Research Centre. The results of these process studies are presented and the implications for water resources are discussed.

Acknowledgements

The research programme described here was carried out under the direction of Dr J.S.G. McCulloch, Director of the Institute of Hydrology, Wallingford, whose support and guidance is gratefully acknowledged.

The members of the Process Interactions Section, whose work is reviewed and reported here, are, in order of their first involvement with the project: Paul Rosier, Philip Hill, Richard Harding, Robin Hall and Ivan Wright.

Thanks are also due to Brian Smith and Celia Kirby for their many helpful comments on the draft, and to my wife, Valerie, for her patient help and assistance in the production of the manuscript.

Symbols

The set of symbols, their definitions and the units in which they are most commonly expressed in this book are given below. The set of symbols chosen is a compromise between consistency, clarity and convenience. The units follow the SI system; where deviations from the SI system occur it is made explicit.

It has been common practice historically in both hydrology and meteorology to express rainfall, evaporation and related terms such as canopy storage as a depth (mm). These terms are also expressed, and it could be argued more properly expressed, as a mass per unit ground area ($kg\ m^{-2}$). The latter definition, although leading to identical numerical results (1 kg of water occupies $0.001\ m^3$), has two advantages: firstly the quantities explicitly refer to ground area (and there can be no confusion with leaf area); secondly there is less possibility of confusion arising from inconsistency of units when evaporation is calculated from equations which include energy terms. Where depth units are used in the text the depth is on a ground area basis unless otherwise stated. Evaporation rates also are expressed in units of energy flux per unit area ($W\ m^{-2}$), representing the energy requirement for the supply of latent heat for evaporation. Pressure is expressed in pascals and, where necessary, figures quoted in earlier literature in bars have been converted ($1\ kPa = 1\ kN\ m^{-2} = 10\ mbar$).

a	annum, year
b	canopy drainage parameter (mm^{-1})
B	snow build up parameter (mm)
c	depth of water retained by vegetation, canopy storage (mm)
c_{max}	maximum depth of water retained by vegetation (mm)
c_L	depth of liquid water retained by intercepted snow (mm)
c_S	water equivalent depth of solid phase water retained by vegetation (mm)

c_T	depth of water retained by vegetation and waxed surface (mm)
c_W	depth of water retained by waxed surface (mm)
c^T	mean total canopy storage value at which the transition in aerodynamic resistance occurs between the sublimation and melt conditions (mm)
c_p	the specific heat of air at constant pressure $(\mathrm{J\ kg^{-1}\ K^{-1}})$
C	heat flux per unit area by convection in air (sensible heat) $(\mathrm{W\ m^{-2}})$; aerodynamic resistance sub-parameter
d	day (86 400 s)
d	zero plane displacement for momentum (m)
d_T	zero plane displacement for temperature (m)
d_q	zero plane displacement for humidity (m)
D	depth rate of drainage from a wet canopy, drainage rate $(\mathrm{mm\ h^{-1}})$
D_T	drainage rate of water on canopy and waxed surface $(\mathrm{mm\ h^{-1}})$
D_W	drainage rate of water on waxed surface $(\mathrm{mm\ h^{-1}})$
$e_s(T)$	saturation vapour pressure of water vapour at temperature T (kPa)
δe	vapour pressure deficit, i.e. $e_s(T) - e$ (kPa)
e_o	vapour pressure of the surface (kPa)
E	water vapour flux per unit area, evaporation rate $(\mathrm{mm\ h^{-1}})$
E^M	measured evaporation rate $(\mathrm{mm\ h^{-1}})$
E_W^{PM}	evaporation rate estimated from the Penman–Monteith equation for a wet surface $(r_s = 0)$ $(\mathrm{mm\ h^{-1}})$
E_T	daily Penman potential transpiration estimate for grass $(\mathrm{mm\ d^{-1}})$
E_{Ta}	annual Penman potential transpiration estimate for grass $(\mathrm{mm\ a^{-1}})$
E_d	daily evaporation estimate $(\mathrm{mm\ d^{-1}})$
E_a	annual evaporation estimate $(\mathrm{mm\ a^{-1}})$
$\bar{E}$	mean evaporation rate $(\mathrm{mm\ h^{-1}})$
f	fraction of the catchment area under complete canopy coverage (dimensionless)
h	hour (3600 s)
h	mean tree height (m)
H	available energy flux per unit area (defined as the net radiation minus the sum of the soil heat and photosynthetic fluxes per unit area) $(\mathrm{W\ m^{-2}})$
i	aerodynamic resistance sub-parameter
I	total depth of intercepted water evaporated per unit ground area, interception loss (mm)
I^M	measured interception loss (mm)

I_1 — unattenuated flux of γ-rays at unit distance from the source (s^{-1})

I_r — attenuated flux of γ-rays at distance r from the source (s^{-1})

k — Von Karman's constant (0.41)

k — canopy drainage parameter in Rutter interception model (mm min^{-1})

k' — canopy drainage parameter in Aston interception model (dimensionless)

K_m — transfer coefficient for momentum ($\text{m}^2\,\text{s}^{-1}$)

L — latent heat of fusion of water (J kg^{-1}); number of elemental surface areas per unit ground area (mm^{-2})

m — mean number of raindrop strikes per element (dimensionless)

M — momentum flux per unit area (N m^{-2}); mass per unit area (kg m^{-2}); melt rate (kg s^{-1})

n — attenuated count rate in γ-ray beam (s^{-1}); mean number of drops retained per element (dimensionless); duration of sunshine hours (h); number of periods (dimensionless)

N — theoretical maximum duration of sunshine hours (h)

n_a — count rate with only air in the γ-ray beam (s^{-1})

n_b — background count rate (s^{-1})

p — total air pressure (kPa); free throughfall fraction (dimensionless)

p_c — maximum liquid to solid ratio of snow (assumed = 0.15) (dimensionless)

P — total depth of precipitation (mm); daily precipitation (mm)

P_a — annual precipitation (mm)

P_r — Poisson probability of an element being struck by r drops (dimensionless)

q — specific humidity (dimensionless); average maximum number of drops that can be retained per elemental area (dimensionless)

$q_s(T)$ — saturated specific humidity at temperature T (dimensionless)

δq — specific humidity deficit, $q_s(T) - q$ (dimensionless)

q_o — specific humidity of the surface (dimensionless)

q_* — humidity scale related to the flux per unit area of water vapour (dimensionless)

Q — net input rate of water to a vegetation canopy (mm h^{-1}); the sum of the rates of precipitation and evaporation (mm h^{-1})

Q_n — evaporation rate equivalent of the net flux of radiation per unit area to the surface (mm d^{-1})

Q_s — net input rate of water to a vegetation canopy in melt

	conditions: the sum of the rates of precipitation evaporation and melt (mm h^{-1})
r	reflection coefficient (dimensionless)
r_a	aerodynamic resistance to the transport of the entities: momentum, heat and vapour, if assumed to be equal (s m^{-1})
r_{aM}	aerodynamic resistance to the transport of momentum (s m^{-1})
r_{aH}	aerodynamic resistance to the transport of heat (s m^{-1})
r_{aV}	aerodynamic resistance to the transport of vapour (s m^{-1})
r_{aL}	aerodynamic resistance to the transport of the entities: momentum, heat and vapour, from a vegetation canopy when retaining liquid phase precipitation (s m^{-1})
r_{aS}	aerodynamic resistance to the transport of the entities: momentum, heat and vapour, from a vegetation canopy when retaining solid phase precipitation (s m^{-1})
r_s	surface resistance to the transport of water vapour (s m^{-1})
R	depth of rainfall per unit time, rainfall rate (s m^{-1})
$\bar{R}$	mean rainfall rate (mm h^{-1})
R_A	evaporation rate equivalent of the incoming radiation flux per unit area at the top of the atmosphere (mm d^{-1})
R_B	evaporation rate equivalent of the net longwave back radiation flux per unit area (mm d^{-1})
R_C	evaporation rate equivalent of the incoming shortwave radiation flux per unit area (mm d^{-1})
R_n	net radiation flux per unit area (W m^{-2})
S	heat flux per unit area by conduction through the surface, soil heat (W m^{-2}); canopy capacity depth Rutter definition (mm)
S_r	sensitivity of γ-ray detector at distance r from the source (dimensionless)
t	time (s); time taken for the central division of a tipping bucket gauge to reach the inlet stream (s)
T	temperature (°C); air temperature (°C); time between successive tips of a tipping bucket gauge
T_o	surface temperature (°C)
T_{oL}	surface temperature calculated on the assumption that the surface is covered with liquid phase water (°C)
T_{oS}	surface temperature calculated on the assumption that the surface is covered with solid phase water (°C)
T_*	temperature scale related to the flux per unit area of heat (°C)

$u(z)$	wind speed at height z above earth's surface (m s^{-1})
u_*	wind scale related to the flux per unit area of momentum, also known as the friction velocity (m s^{-1})
v	mean volume of raindrops (m^3)
V	static volume of tipping bucket gauge (dm^3)
w	fraction of the day that the vegetation canopy is wet (dimensionless)
w_a	fraction of the year that the vegetation canopy is wet (dimensionless)
z	height above the earth's surface (m)
z_o	roughness length for momentum (m)
z_T	roughness length for temperature (m)
z_q	roughness length for humidity (m)
α	interception fraction, proportion of precipitation lost by interception (dimensionless)
β	parameter relating daily transpiration to the Penman E_T estimate for grass (dimensionless)
γ	psychrometric 'constant' (kPa $^\circ$C^{-1}); daily interception model parameter (mm); γ-ray
Δ	rate of change of saturation vapour pressure with temperature (kPa $^\circ$C^{-1}); finite difference increment
Δ'	rate of change of saturated specific humidity with temperature ($^\circ$C^{-1})
δ	daily interception model parameter (mm^{-1})
ϵ	ratio of molecular weight of water and air (0.622)
λ	latent heat of vaporisation of water (J kg^{-1})
μ	attenuation coefficient for γ-rays in a homogeneous absorber (m^2 kg^{-1})
π	ratio of the circumference to the diameter of a circle (3.14159265359)
ρ	density (kg m^{-3}); density of moist air (kg m^{-3})
ρ_a	density of dry air (kg m^{-3})
Σ	sum of a series
σ	Stefan–Boltzmann constant (5.67×10^{-8} W m^{-2} K^{-4})
τ	temperature-dependent regulating function for grass evaporation (dimensionless)

Prefixes

μ	micro (10^{-6})
m	milli (10^{-3})
c	centi (10^{-2})
d	deci (10^{-1})
k	kilo (10^3)
M	mega (10^6)

CHAPTER 1

Introduction

1.1 FRANK LAW AND THE STOCKS RESERVOIR CONTROVERSY

Frank Law, engineer to the Fylde Water Board, presented, in 1956, results from his hydrological studies in the Lancashire Pennines. These studies indicated that the evaporative losses from coniferous forest were greater than from grassland and that, to the water engineer, the magnitude was such that any economic return from the afforestation was far outweighed by the value of the lost water. At the time, these results were greeted with scepticism; it was then generally believed that evaporation from natural catchments was closely linked to the availability of (solar) radiant energy and that the evaporation from all crops, whether forest or grass, (if adequately supplied with water and subject to the same climate) would be broadly similar. Indeed, it was argued on occasions that (although no experimental evidence exists in the UK) forests, particularly tropical forests, encourage rainfall and that runoff from forested catchments may actually be enhanced. Doubts also arose because of the small size of the forest used by Law as his study area: because of this the possibility of aerodynamic and radiative edge effects could not be discounted. Despite these reservations it was generally recognised that the questions posed by Law's results were of sufficient gravity to warrant further study.[1]

Subsequent research has shown that Law's interpretation of his results was correct. His observations have been verified elsewhere in the British uplands and there is now a much greater understanding of the processes which, in wet climates, enhance evaporative losses from forests as compared with shorter upland vegetation types.

The studies which led to this new understanding are discussed in this chapter. The remaining controversy concerning the

---

[1]The Hydrology Research Unit was formed in 1968 to investigate this and other topics and the research theme continued when the Unit was later incorporated within the Natural Environment Research Council as the Institute of Hydrology.

differences in evaporation between forest and upland vegetation
types other than grass, and the differences that may occur on a
seasonal basis, are treated in later chapters.

1.2 FRANK LAW'S STUDIES

Frank Law began his hydrological studies at Stocks Reservoir in
1954. The reservoir, located in the Lancashire Pennines, has a

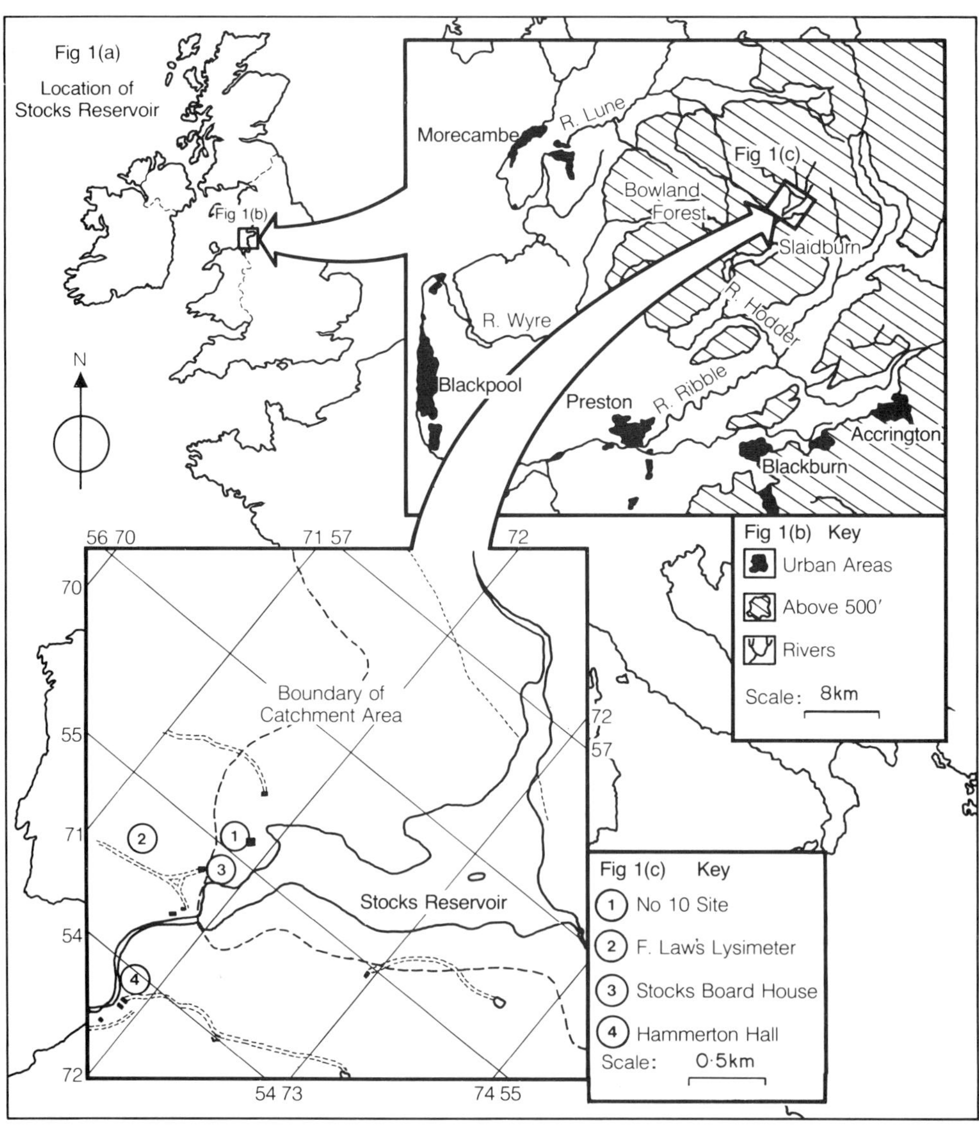

Figure 1.1 Location of hydrological experiments at Stocks Reservoir.

catchment area of 37.5 km^2 and lies at an altitude between 180 and 540 metres above sea level receiving an average annual rainfall of 1650 mm (Figure 1.1). Data were available from the early 1930s of reservoir abstractions, outflow and change of reservoir storage, allowing runoff into the reservoir to be calculated. Rainfall measurements were available from five gauges dating from 1910, and during Law's term of office the number was increased to 22. Runoff measurements from a neighbouring non-afforested catchment, Croasdale, were initiated in 1957 and in 1960 a flume was constructed to measure flows from Bottoms Beck, the main afforested sub-catchment of the reservoir. A second full climatological station was constructed on an exposed promontory near the reservoir to house grass and heather weighing lysimeters and a range of experimental raingauges. A forest lysimeter was also constructed occupying 450 m^2 of mature sitka spruce (*Picea sitchensis*) plantation (Figure 1.2) and throughfall and stemflow measurements (Figure 1.3) were made within the surrounding forest plantation.

Law was quick to appreciate the significance of the results he obtained from these experiments and in 1956 (Law, 1956, 1957) he published his finding that 290 mm less rainfall was converted to runoff on the forest lysimeter; he concluded that if the whole catchment was afforested the effect would be to reduce water supplies by 42%. Measurements were continued at Stocks until 1970 but thereafter Law did not formally publish any of these

Figure 1.2 Law's sitka spruce lysimeter. (The photograph was taken in 1980, ten years after measurements from the lysimeter ceased.)

Table 1.1 Summary of data from Stocks Reservoir—sitka lysimeter and catchment areas (units: mm). (Reproduced by permission from Calder *et al.*, 1982b.)

	Calendar years															Averages	
	1956	1957	1958	1959	1960	1961	1962	1963	1964	1965	1966	1967	1968	1969	1970	All years shown	6-year: 1961–64, 1966, 67
SITKA PLANTATION LYSIMETER (450 m^2)																	
1. Rainfall: mean of 3 outside gauges	1374	1557	1491	1245	1588	1615	1468	1367	1323	1552	1651	1915	1547	1184	1565	1496	1556
2. Throughfall: stemflow and plantation gauges	871	1003	1039	757	1001	993	902	836	828	955	1008	1062	1072	678	897	927	938
3. Interception: (1−2)	503	554	452	488	587	622	566	531	495	597	643	853	475	506	668	569	618
4. Lysimeter runoff	411*	734	762	447	523	538	531	457	510	†	531	663	†	†	†	555	538
5. Gross losses: (1−4)	963	823	729	798	1065	1077	937	910	813	−	1120	1252	−	−	−	953	1018
6. Net (evapotranspiration) losses: (2−4)	460	269	277	310	478	455	371	379	318	−	477	599	−	−	−	381	400
STOCKS RESERVOIR CATCHMENT (22% afforested) (37.5 km^2, including Bottoms Beck)																	
7. Estimated area rainfall	1471	1725	1665	1370	1785	1811	1640	1519	1412	1740	1841	2133	1737	1328	1750	1662	1726
8. Catchment runoff	1038	1208	1200	907	1375	1439	1212	1143	979	1270	1355	1565	1289	881	1254	1208	1282
9. Losses: (7−8)	433	517	465	463	410	372	428	376	433	470	486	568	448	447	496	454	444
BOTTOMS BECK CATCHMENT (70% afforested) (10.6 km^2)																	
10. Estimated areal rainfall					1674	1458	1405	1269	1575	1684	2065	1593	1224	1494		1544	1592
11. Catchment runoff					1250	1016	993	808	1092	1191	1341	1123	734	945		1049	1100
12. Losses (10−11)					414	442	412	460	483	493	724	470	490	549		495	492

*Not recalculated from original field sheets.

†Runoff data thought to be unreliable for all or major part of year.

21 trees removed from lysimeter in March 1962 and a further 17 in December 1967 leaving 57 thereafter.

In May 1957 there was a severe loss of needles from trees in the lysimeter due to an aphid attack.

Figure 1.3 Stemflow gauge on a tree adjacent to Law's lysimeter. (The photograph was taken in 1980, ten years after measurements from the lysimeter ceased.)

results to substantiate his earlier key findings. This mass of data has proved to be of immense value in understanding the effects of afforestation in the uplands (Walsh, 1977, 1980; Calder *et al.*, 1982b; Walsh and Walker, 1985) and substantiates Law's initial findings; however, recalculation of the areal rainfall values reveals losses even greater than those originally quoted by Law. A summary of the lysimeter and catchment data for the years 1956–1970 is given in Table 1.1.

1.3 THE PLYNLIMON CATCHMENT RESULTS

To investigate Law's claims further, the newly formed Hydrological Research Unit at Wallingford initiated a paired catchment experiment to compare and contrast the hydrological regimes

Table 1.2 Annual values of precipitation (P), runoff (Q), losses ($P-Q$) and difference in losses for the grassland Wye catchment and the forested area of the Severn catchment only, for the years 1970–80; units: mm. (Reproduced by permission from Calder *et al.*, 1982b.)

Year	Precipitation (P)		Runoff (Q)		Loss ($P-Q$)		Difference in losses
	Wye	Severn	Wye	Severn	Wye	Severn	(Severn−Wye)
1970	2869	2485	2415	1636	454	349	+395
1971	1993	1762	1562	797	431	965	+534
1972	2131	2124	1804	1342	328	782	+454
1973	2606	2380	2164	1581	442	799	+357
1974	2794	2703	2320	1785	474	918	+444
1975	2099	2035	1643	1213	456	822	+366
1976	1736	1645	1404	921	332	724	+392
1977	2561	2573	2236	1638	325	935	+610
1978*	2356	2367	2128	1668	228	699	+471
1979*	2742	2683	2463	2016	279	667	+388
1980	2695	2517	2377	1914	318	603	+285
Mean	2417	2300	2047	1501	370	799	+429

*Years in which snow made necessary the estimation of monthly precipitation total in winter.

Figure 1.4 Headwaters of the River Severn at Plynlimon, central Wales.

associated with upland grass moorland and coniferous afforestation. The area chosen was the headwaters of the Wye and Severn
catchments at Plynlimon in mid-Wales (Figure 1.4) and the scale
of the experiments was such that the possibility of edge effects,
which were thought to be associated with Law's lysimeter, could
be discounted. By 1970 continuous records were being obtained
from the catchments. Table 1.2 shows that after adjustment for
the unforested parts of the Severn catchment, which are assumed
to have the same rainfall-runoff relationship as the Wye, losses
from a catchment covered with mature forest would be twice that
from grass moorland.

1.4 THE PLYNLIMON FOREST LYSIMETER

Within the overall framework of the catchment experiments at
Plynlimon, 'process' studies were carried out to obtain a better
understanding of the mechanisms which control evaporation from
upland catchments and so allow extrapolation of the Plynlimon
results to other areas.

   One of these studies involved the construction of a forest
lysimeter within the forested Severn catchment (Figures 1.5 and

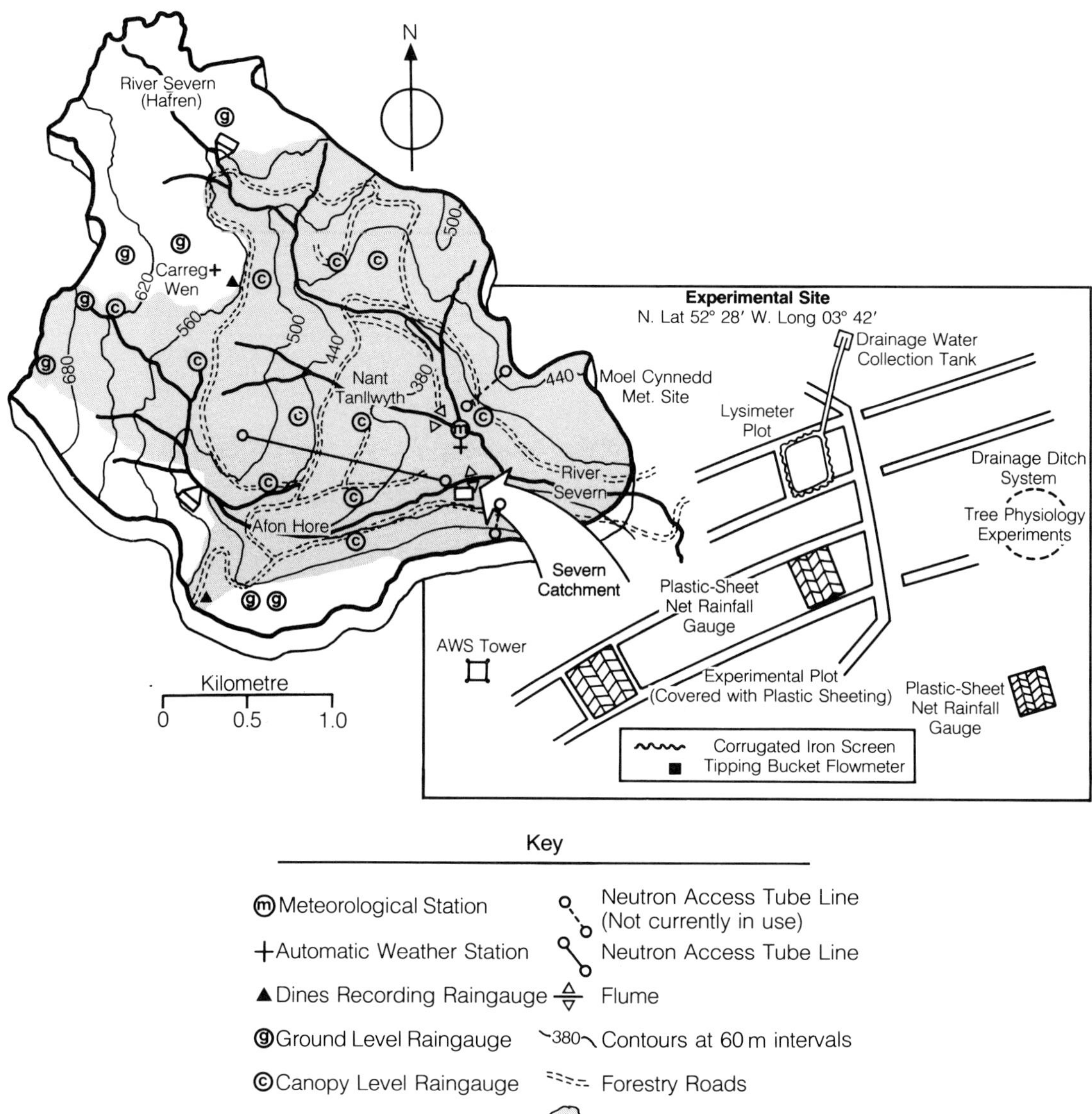

Figure 1.5 Location of hydrological experiments in the Hafren Forest, Plynlimon.

1.6), which was, in principle, similar to Law's lysimeter at Stocks. A corrugated iron screen was installed into clay subsoil to isolate hydraulically an 84 m² plot which enclosed 26 spruce trees. The forest stand, which was planted in 1937, consisted mainly of Norway spruce (*Picea abies* (L.) Karsten) with a small admixture of sitka (*Picea sitchensis* (Bong.) Carrière); the stocking density was 3200 stems ha⁻¹ and the mean tree height was 12 m.

Figure 1.6 A neutron soil-moisture probe being used on the Plynlimon forest lysimeter.

However, unlike the Stocks lysimeter, a neutron probe was used to measure soil water content and so enable the water balance of the lysimeter to be calculated over much shorter time intervals. The neutron probe measurements also provided a check on the basic water balance measurements of rainfall, net-rainfall and drainage, and confirmed that leakage losses from the lysimeter were negligible (Calder, 1976, 1985b). Rainfall and meteorological measurements were made in a clearing 400 m distant from the site; net-rainfall measurements were obtained from a replicate pair of large plastic-sheet net-rainfall gauges (Calder and Rosier, 1976) on a plot adjacent to the lysimeter. The lysimeter was operated from February 1974 to October 1976 and as a 'covered plot' experiment (Figure 1.7) through the winter of 1976–77 (Calder, 1976, 1977, 1978); physiological studies were also carried out during 1975 at the site (Roberts, 1978). The results from the lysimeter clearly demonstrated the source of the high losses from forests in wet upland climates: interception losses (the difference between recorded rainfall and net-rainfall) were, on an annual basis, almost twice those due to transpiration (Figure 1.8, Table 1.3) and in total amounted to almost twice the Penman (1948) potential transpiration estimate (E_T) for grass. On a long-term basis (annual) the total transpiration on dry days was found to be 0.9 of the Penman E_T total.

Figure 1.7 Soil moisture measurements being taken beneath the covered lysimeter during the winter of 1976–77 to determine winter transpiration losses.

The detailed measurements from an automatic weather station (Strangeways, 1972) of rainfall and net-rainfall and other meteorological variables on a five-minute basis allowed the parameters in a Rutter-type interception model (see section 3.2) to be

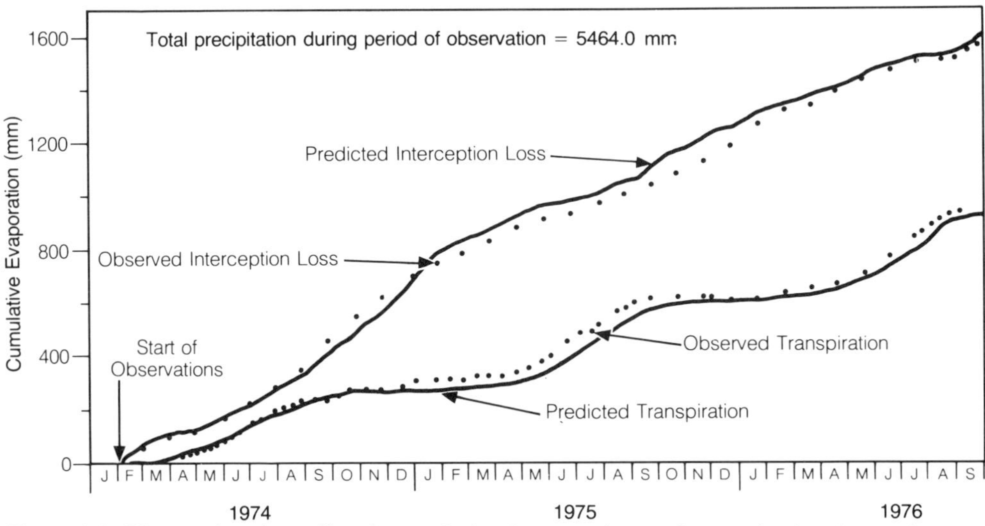

Figure 1.8 Observed and predicted cumulative interception and transpiration losses from the Plynlimon forest lysimeter. (Reproduced by permission from Calder, 1979a; Copyright Water Services.)

Table 1.3 Evaporation measurements from the Plynlimon forest lysimeter, February 1974–September 1976; units: mm. (Reproduced by permission from Calder *et al.*, 1982b.)

Period	Precipitation	Interception	Transpiration
6 February 1974–31 December 1974	3238	685	289
1 January 1975–31 December 1975	2013	529	335
1 January 1976–1 October 1976	1103	366	277
TOTAL	5444	1580	901

determined by optimization procedures. The mean parameter values were found to be:

p = 0.005, ('free throughfall' fraction)
k = 0.00028 (drainage parameter, mm min^{-1})
b = 1.76 (drainage parameter, mm^{-1})
s = 0.37 (canopy storage capacity, mm).

Similar optimization methods were used, in conjuction with the daily estimates (in summer) of transpiration from the lysimeter, to develop and calibrate a surface resistance sub-model which related the hourly surface resistance (r_s) to the day number of the year (D) and the hourly value of the vapour pressure deficit δe (kPa). The functional form was:

$$r_s = 74.5(1-0.30\cos(2\pi(D-222)/365))/(1-0.45\delta e) \quad \delta e<2.2; \quad 1.1$$
$$r_s = \infty, \qquad\qquad\qquad\qquad\qquad\qquad \delta e\geq2.2$$

The predictions obtained with this model are shown in Figure 1.8 together with observations.

The parameter values which related the surface resistance to vapour pressure deficit were also verified using tree cutting and diffusion porometry methods on trees near the lysimeter (Roberts, 1978). A notable 'non-result' from these studies was the lack of any evidence, even during the exceptionally dry summer of 1976 when soil moisture deficits exceeding 200 mm were recorded (equivalent to a soil water tension of 600 kPa), to suggest that soil moisture limitations were restricting evaporation rates.

1.5 THE THETFORD MICROMETEOROLOGICAL AND SOIL PHYSICAL STUDIES

In conjunction with the studies carried out at Plynlimon the Institute of Hydrology initiated an experiment (Figure 1.9) to study the evaporation from a forest using measurements of the temperature, humidity and wind profiles above the canopy and

Figure 1.9 The Thetford micrometeorological studies.

measurements of the energy balance. A requirement of the method was an extensive area of flat forest which would allow full development of the profiles without introducing the additional complications of flux divergence, which may be brought about by either unevenness in the surface or discontinuities in vegetation type. Various methods were used to measure evaporation rates including the Bowen ratio method, using both fixed and interchanging sensors (Stewart and Thom, 1973) and, more recently, eddy correlation devices. Soil physical studies (Cooper, 1980) and physiological studies were also carried out at the site. The results from these different studies were generally internally consistent and demonstrated that transpiration rates, although on occasions strongly influenced by stomatal control mechanisms, were broadly similar to the Penman E_T estimate. In wet conditions large differences were found: evaporation rates were recorded which greatly exceeded those that could be sustained by radiant energy inputs (Stewart, 1977).

Figure 1.10 Predicted percentage increase in evaporative loss as a result of afforesting to 75% forest cover (equivalent to 50% canopy cover) the catchments supplying the major British upland reservoirs.

1.6 THE CONTINUING CONTROVERSY

The studies outlined above all confirm Law's original findings (see Binns, 1980): although when well supplied with water grass may transpire more than forest, in wet conditions evaporation rates of intercepted water will be greatly enhanced from the taller vegetation and, in the wet upland areas of the UK, will totally

Figure 1.11 Upland catchments on which an expansion of forest cover would affect water supply or power generation.

dominate any small differences in transpiration. Calder and Newson (1979) suggested a simple semi-empirical method (see section 9.3) to calculate the effects (on an annual basis) of afforesting grass catchments in the uplands, based essentially on the observations that interception losses from forests are about 35% of the annual rainfall; transpiration from forest is at about the E_T rate, whilst from grass both transpiration and interception occur at about the E_T rate. This simple model was used to investigate the effects on water supplies of afforesting the catchments of the major UK reservoirs (Figure 1.10) (Calder and Newson, 1979, 1980) and to provide information for the Centre for Agricultural Strategy's investigations (CAS, 1980) into the feasibility of proposals to increase greatly the proportion of forestry in the British uplands.

The model as presented was only strictly applicable to situations where afforestation was taking place on previously grass moorland and in areas where rain formed the major part of the annual precipitation. These limitations, and in particular how these limitations might affect the estimates for the Scottish uplands, were discussed in public debate and it was recognized that the results were of sufficient importance to the water industry (see Figure 1.11) for further studies to be carried out.

These studies, funded by a consortium comprising the British Waterways Board, Department of Energy, North of Scotland Hydro-electricity Board, Scottish Development Department, Forestry Commission, and a number of regional councils and water authorities represented by the Water Research Centre, aimed at further quantification of the evaporation rates from the major upland vegetation types in dry, rain and snow conditions. For the water supply interests, the vital question of the seasonal differences in evaporation loss between the upland vegetation types was also investigated.

CHAPTER 2

Physics and Physiology of Evaporation

2.1 INTRODUCTION

Evaporation in the natural environment can be considered to be the result of a balance between externally applied atmospheric demand and the availability of water at the evaporating surface. The influence of these variables on evaporation rates is confused because to a greater or lesser extent both variables are themselves determined by the evaporation rates achieved. Central to present research into evaporation is the study of these feedback processes.

Theories of evaporation can be traced to Aristotle's qualitative observations that both the sun's heat and wind were important in determining evaporation, early recognition of the controlling thermodynamic and aerodynamic factors. Dalton's experiments in the early nineteenth century provided the means for quantifying evaporation rates using the aerodynamic approach. The concept of the energy balance at the evaporating surface followed and was combined with the aerodynamic approach through the use of the 'del' approximation (see section 2.4). Further developments led to the explicit treatment of the availability of water at vegetative surfaces through the introduction of the concept of a resistance imposed by the vegetation on the movement of water to the outer surface of the vegetation, termed the 'surface resistance'. Later studies attempted to relate surface resistance to other physiological and environmental factors, including atmospheric demand and soil water status. The transport of heat, vapour and momentum from the evaporating surface were conventionally described using one-dimensional models of turbulent diffusion; logarithmic profiles of the concentrations of these quantities are predicted which are generally confirmed by observation. However, other processes are also at work. Transport by gust penetration is likely to be significant from canopies with open structures which allow horizontal movement of air. Another, equally poorly understood mechanism is the feedback effect of evaporation on the atmosphere.

The development of these ideas and the way in which they

have been formulated into 'research' models is outlined in this chapter. A generalised model of evaporation from natural systems has not been achieved; yet even if such a model was possible, limitations of input data would almost certainly make it unusable. An important aspect of current applied research is to determine which approximations, given a particular objective, are possible so that practical evaporation equations can be developed which have only minimal data requirements. The loss of model generality necessary with these models need not also result in a loss of model accuracy, indeed it may often be possible for simple methods, over the range of conditions for which they were designed, to provide more accurate results than their more sophisticated counterparts. The development of these methods is outlined in Chapter 9.

2.2 THE DALTON EQUATION

Dalton (1801) showed that evaporation rates from water warmed in small containers were closely related to the difference between the vapour pressure of water in the container and the vapour pressure of water at the dew point temperature of the air. He was also aware of the importance of air circulation in promoting evaporation, and these observations led to the 'aerodynamic' or 'Dalton' equation:

$$E = (e_o - e)\, f(u)$$

where e_o is the vapour pressure of the surface, e is the vapour pressure of the air and $f(u)$ is some function of windspeed.

If the aerodynamic function is expressed as a resistance (r_a) to the transfer of water vapour down the vapour concentration gradient that exists between the evaporating surface and the atmosphere (Figure 2.1) and the evaporation rate is expressed as the equivalent rate of latent heat, λE, the (more exact) Dalton equation becomes:

$$\lambda E = \lambda \rho_a \epsilon (e_o/(p-e_o) - e/(p-e))/r_a$$

where ρ_a is the density of dry air, ϵ is the ratio of the molecular weights of water and air and p is the total air pressure. On substitution of the density of dry air for the density of moist air using the relation, $\rho_a = \rho(p-e)/(p-(1-\epsilon)e)$, gives:

$$\lambda E = \lambda \rho \epsilon (e_o/(p-(1-\epsilon)e_o) - e/(p-(1-\epsilon)e))/r_a$$

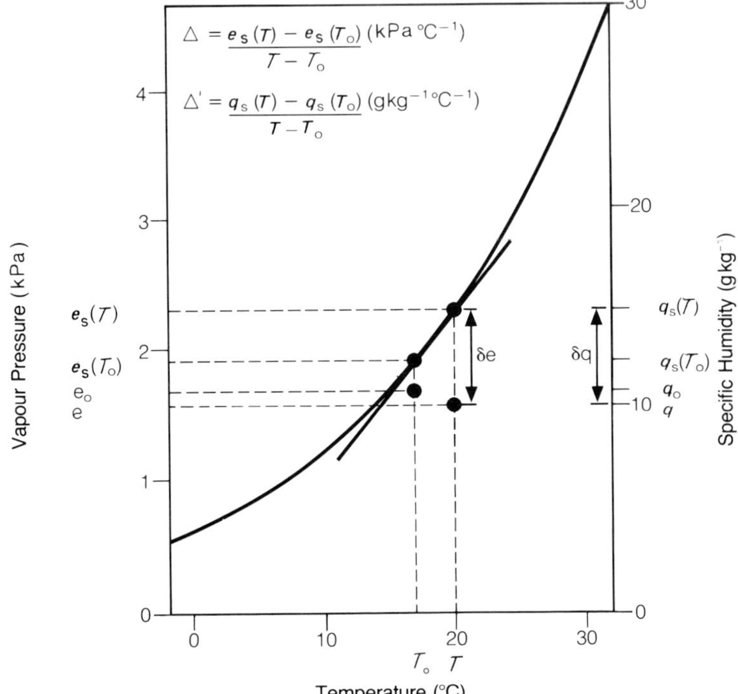

Figure 2.1 Relationship between saturated vapour pressure (or saturated specific humidity) and temperature illustrating the 'del' approximation.

As the total air pressure, p, is usually much larger than the terms $(1-\epsilon)e$ and $(1-\epsilon)e_o$, the equation can be simplified with little loss of accuracy (typically of the order of 1%) and is usually expressed, after rearrangement of terms, as:

$$\lambda E = \frac{\rho c_p}{\gamma}(e_o - e)/r_a \qquad\qquad 2.1$$

where c_p is the specific heat of air and the group of terms, $c_p p/\lambda\epsilon$ is termed the psychrometric 'constant', γ, and has the value of 0.066 kPa °C^{-1} at a temperature of 20 °C and a pressure of 100 kPa (McCulloch, 1965, see also Table A.1).

For completeness, the development of the combination equation in terms of specific humidity, which avoids the use of the above approximations, is also given.

Specific humidity (q), defined as the mass of water vapour contained per unit mass of moist air (kg kg^{-1}; g kg^{-1}), is related to vapour pressure via the relation:

$$q = \frac{\epsilon e}{p - (1 - \epsilon)e}$$

The equivalent of equation 2.1, expressed in terms of specific humidity, is therefore given by:

$$\lambda E = \lambda\rho(q_o - q)/r_a \qquad\qquad 2.1a$$

where ρ is the density of moist air and q_o and q are the specific humidities of the surface and air respectively.

Sensible heat (C) may also be described in terms of the similar relationship:

$$C = \rho c_p(T_o - T)/r_a \qquad\qquad 2.2$$

where T_o and T are the surface and air temperatures respectively.

2.3. THE ENERGY BALANCE EQUATION

Evaporation rates from vegetation can be calculated from solution of the energy balance equation:

$$R_n = \lambda E + C + S + p$$

where R_n is the net radiation input, S is the soil heat flux and p is the energy used in photosynthesis. The photosynthesis term is small compared with the other terms and can generally be neglected, as can the soil heat flux term, when evaporation rates are calculated over long time periods (exceeding 1 day). If available energy (H) is defined as the net radiation minus the sum of the soil heat flux and photosynthetic terms, the energy balance equation may be rewritten as:

$$H = \lambda E + C \qquad\qquad 2.3$$

2.4 THE 'DEL' APPROXIMATION AND THE COMBINATION EQUATION

In practice it is not usually possible to achieve an independent solution of either the aerodynamic or energy balance equations; although the meteorological variables—net radiation, air temperature and humidity (R_n, T, e or q)—are relatively easy to measure and the $f(u)$ or r_a relationship can (in principle) be determined, the surface temperature and humidity terms (T_o, e_o or q_o) and the sensible heat term (C) are generally unknown. Nor is the solution of the three equations as a set of simultaneous equations possible because, including the latent heat term, there are four unknowns. The 'del' approximation, used by Penman (1948) in the derivation of the Penman equation provides a possible solution; this defines a fourth equation through knowledge of the change with temperature of the saturated vapour pressure of water (Figure 2.1), whilst introducing no further unknown terms:

$$\Delta = \frac{e_s(T) - e_s(T_o)}{T - T_o} \qquad 2.4$$

where e_s represents the saturated vapour pressure at the temperature given in parentheses.

In terms of specific humidity the equivalent relationship is:

$$\Delta' = \frac{q_s(T) - q_s(T_o)}{T - T_o} \qquad 2.4a$$

If the surface of the vegetation is wet, the humidity at the surface expressed as a vapour pressure is given by:

$$e_o = e_s(T_o) \qquad 2.5$$

and expressed as a specific humidity:

$$q_o = q_s(T_o) \qquad 2.5a$$

Elimination of the surface temperature, surface humidity and sensible heat terms (T_o, e_o, C) from equations 2.1, 2.2, 2.3, 2.4 and 2.5 gives the combination equation in terms of vapour pressure:

$$\lambda E = \frac{\Delta H + \rho c_p(e_s(T) - e)/r_a}{\Delta + \gamma} \qquad 2.6$$

Elimination of the surface temperature, surface humidity and sensible heat terms (T_o, e_o, C) from equations 2.1a, 2.2, 2.3, 2.4a and 2.5a gives the combination equation in terms of specific humidity:

$$\lambda E = \frac{\Delta' H + \rho c_p(q_s(T) - q)/r_a}{\Delta' + c_p/\lambda} \qquad 2.6a$$

The terms $(e_s(T) - e)$ and $(q_s(T) - q)$, which relate to the saturation deficit of the atmosphere, are referred to as the vapour pressure deficit, δe, and the specific humidity deficit, δq, respectively.

For convenience Δ or Δ' is usually calculated at air temperature. When the difference between surface and air temperatures is only a few degrees Celsius, which is commonly the case when the surface is either wet or transpiring vegetation, the error that is introduced by this approximation is insignificant compared with other experimental errors of measurement. In certain circumstances, for example a saline lake (Calder and Neal, 1984) or from a bare earth surface, the temperature difference may be of the order of tens of degrees Celsius and there may be some justification for calculating a better approximation to Δ or Δ' by using an initial trial value calculated at the air temperature and then using an iterative procedure using successive estimates of the surface temperature to calculate Δ from equation 2.4 or Δ' from equation 2.4a.

It is instructive to consider the combination equation to be the sum of two components: the radiative term, $\Delta H/(\Delta + \gamma)$, which, in units of specific humidity, is given by $\Delta'H/(\Delta' + c_p/\lambda)$; and the aerodynamic term, $(\rho c_p\, \delta e/r_a)/(\Delta + \gamma)$ which, in units of specific humidity, is given by $(\rho c_p \delta q/r_a)/(\Delta' + c_p/\lambda)$. In a situation where evaporation in taking place into a confined space or 'closed box' the aerodynamic term would eventually tend to zero and the evaporation would tend to the rate, sometimes termed the equilibrium rate, given by the radiative term. On the other hand, in conditions of zero net radiation the evaporation rate will be determined solely by the aerodynamic or, as it is sometimes referred to, the advective term, in the equation. The value of the aerodynamic resistance is critical in determining the relative importance of the two components; generally we expect that for short vegetation, with a high aerodynamic resistance, evaporation rates will be controlled by the radiative term (i.e. close to the equilibrium rate), whereas for taller vegetation the advective term will assume greater importance.

2.5 THE PENMAN EQUATION

The equation originally derived by Penman (1948), although based on the physical reasoning outlined in the previous section, incorporated certain empiricisms to take account of the aerodynamic and net radiation terms. For an open water surface the evaporation rate, E_o (mm d^{-1}) was given by:

$$E_o = \frac{\Delta Q_n + \gamma E_a}{\Delta + \gamma} \qquad 2.7$$

where the evaporation equivalent of the available energy, Q_n (mm d^{-1}) was calculated from the evaporation equivalents of the incoming shortwave radiation, R_C (mm d^{-1}), minus the fraction, r, reflected, minus the net longwave back radiation from the earth, R_B (mm d^{-1}), i.e.:

$$Q_n = R_C(1 - r) - R_B \qquad 2.8$$

For water the reflection term, r, was taken to be 0.05.

The term R_C was estimated from an empirical relationship based on the duration of bright sunshine hours, n, the maximum possible duration of sunshine hours, N, and the theoretical amount of radiation, R_A (mm d^{-1}), received at the top of the atmosphere:

$$R_C = R_A(0.18 + 0.55n/N) \qquad 2.9$$

Penman calculated the net longwave back radiation by using the equation for calculating the back radiation into a cloudless sky (originally proposed by Angstrom (1920), quoted by Brunt (1939)) together with an empirical term to take account of the cloudiness:

$$R_B = 86400\sigma T^4(0.56 - 0.248e^{0.5})(0.10 + 0.90\,n/N)/\lambda \qquad 2.10$$

where, in SI units, T is the mean air temperature (K) and e is the water vapour pressure in the air in units of kPa.

The aerodynamic term was calculated from the empirical relationship:

$$E_a = 2.6(e_s(T) - e)(1 + 0.537u_2) \qquad 2.11$$

where u_2 is the average daily wind speed at a height of 2 m (m s^{-1}) and vapour pressure is in units of kPa.

Penman found that the evaporation from short vegetation (grass) when well supplied with water, which he termed the 'potential' transpiration, E_T, was related to the open water estimate via the relationship:

$$E_T = fE_o \qquad 2.12$$

where the factor f was 0.8 in summer, 0.7 in spring and autumn and 0.6 in winter.

Later, Rijtema (1965) showed that similar E_T values could be obtained directly, without the use of the f factor, by incorporating in equations 2.8 and 2.7 the reflection coefficient for a grass surface ($r = 0.25$) rather than that for a water surface ($r = 0.05$), ie:

$$E_T = \frac{\Delta Q_n' + \gamma E_a}{\Delta + \gamma} \qquad 2.13$$

where Q_n' is now the evaporation equivalent of the available energy over a grass surface and is given by:

$$Q_n' = (1 - 0.25)R_A(0.18 + 0.55n/N)$$
$$- 86400\sigma T^4(0.56 - 0.248e^{0.5})(0.10 + 0.90\,n/N)/\lambda$$

2.14

Later workers have sometimes used different formulations of the empirical relationships and different empirical factors and this has introduced a certain degree of uncertainty into the exact definition of the Penman equation. The formulation that has been used in this text, and which is referred to as the Penman E_T potential transpiration estimate for short grass and is used essentially as a reference index of evaporation, is that given by equation 2.13.

This approach to estimating evaporation developed by Penman, and independently by Ferguson (1952), was recognised as having a sound theoretical basis, gained worldwide acceptance, and is now perhaps the most commonly used method for estimating evaporation in the environment. Estimates of Penman evaporation have been calculated for most countries in the world and a vast body of experience now exists, exemplified by the FAO Report No. 24, *Irrigation and Drainage Paper*, (Doorenbros and Pruitt, 1977), which allows the calculation of actual evaporation from crops by application of different 'crop factors' to the Penman estimate.

The validity of the equation has been questioned in certain applications, however, and its relevance in the research field is certainly now diminishing. As pointed out by Thom and Oliver (1977), equation 2.11 underestimates the aerodynamic term for the vast majority of applications and because surface factors (stomatal resistance, see next section) were ignored by Penman, the radiative term is usually effectively overestimated. In some circumstances, for example when applied to grass vegetation, these two errors are mutually compensating but in certain applications large errors can result.

Thom and Oliver showed that the roughness length (see section 2.8) implicit in the Penman aerodynamic term, equation 2.11, is about 1.4 mm, which would only be appropriate for very smooth surfaces, for example very short grass. For regional applications roughness lengths of the order of at least a few centimetres would be appropriate, whereas for forests roughness lengths would be of the order of metres.

To correct these omissions Thom and Oliver suggested the incorporation of two more parameters, m and n, in the Penman equation to take into account the different surface roughnesses and different surface factors. The Thom and Oliver version of the Penman equation then becomes:

$$E_T = \frac{\Delta Q_n + m\gamma E_a}{\Delta + \gamma(1+n)}$$

2.15

They suggested regional values, for the representative county of Buckinghamshire, of 2.5 for m and 1.4 for n. By analogy with the Penman–Monteith equation (equation 2.18) the parameter n can be interpreted as the ratio of the surface to aerodynamic resistances.

Attempts have also been made to simplify the Penman equation for use in situations where data are limited. Linacre (1977) removed the need for measurements of radiation and vapour pressure by empirically relating these terms to mean temperature, temperature range, altitude and latitude.

Another widely used equation was developed by Priestley and Taylor (1972) for use over land sites which were neither suffering from water stress nor experiencing significant advection. Essentially their equation expresses the advective term as a constant fraction, via the parameter α, of the equilibrium rate, ie.:

$$\lambda E = \alpha(\Delta/(\Delta + \gamma))(R_n - S) \qquad 2.16$$

For many sets of observations on short agricultural crops Priestley and Taylor found that the empirical factor, α, was close to 1.26.

For the Penman equation and its derivatives, it is of the utmost importance that the assumptions and empiricisms inherent in the equations are understood so that the equations are not used outside their range of applicability. The concern expressed by Thom and Oliver (1977) and Oliver (1983) in applying the Penman aerodynamic term to tall crops is fully borne out by observation. Annual evaporation rates have been measured from forests growing in wet upland areas (Calder, 1976) which exceed the Penman E_o estimate by a factor of two. Shuttleworth and Calder (1979) have also shown, for the same reasons, that the Priestley and Taylor form of the equation is equally inappropriate in these circumstances.

In view of the above discussion it is somewhat surprising that the transpiration rates from spruce forest in dry conditions are found to be quite close to the Penman potential transpiration values. This apparent paradox can be explained in terms of the physiological response mechanism which some crops exhibit to changes in atmospheric humidity deficit. For spruce forest it is found that as atmospheric humidity deficits increase there is an increase in stomatal resistance, which makes the transpiration rate relatively insensitive to changes in humidity. Somewhat fortuitously, although the equation does not explicitly take into account this physiological response mechanism, the Penman equation, by underestimating the aerodynamic term, does nevertheless give reasonably good estimates of spruce transpiration in dry conditions—usually within 10% of, and usually 10% less than, long-term observations.

However, the evaporation rates in wet conditions differ so widely from the E_T value that it is now usual, for practical purposes, to treat them separately with an empirical relationship based on rainfall amount (see section 9.3).

Although the Penman equation cannot now be thought of as providing an estimate of 'potential' evaporation the wide availability of Penman estimates in time and space will ensure the equation's continued application as a 'reference' in estimating evaporation from both short and tall crops for some time to come.

2.6 PHYSIOLOGICAL RESISTANCE—THE PENMAN–MONTEITH EQUATION

Monteith (1965) introduced another parameter, r_s, into the combination equation to deal explicitly with the physiological resistance that crops impose on the transfer of water from within their internal organs to their outer surfaces (Figure 2.2). This resistance is defined by the equation:

$$\lambda E = \frac{\rho c_p}{\gamma}(e_s(T_o) - e_o)/r_s \qquad 2.17$$

or, in terms of specific humidity, the equivalent relationship is:

$$\lambda E = \lambda \rho C(q_s(T_o) - q_o)/r_s \qquad 2.17a$$

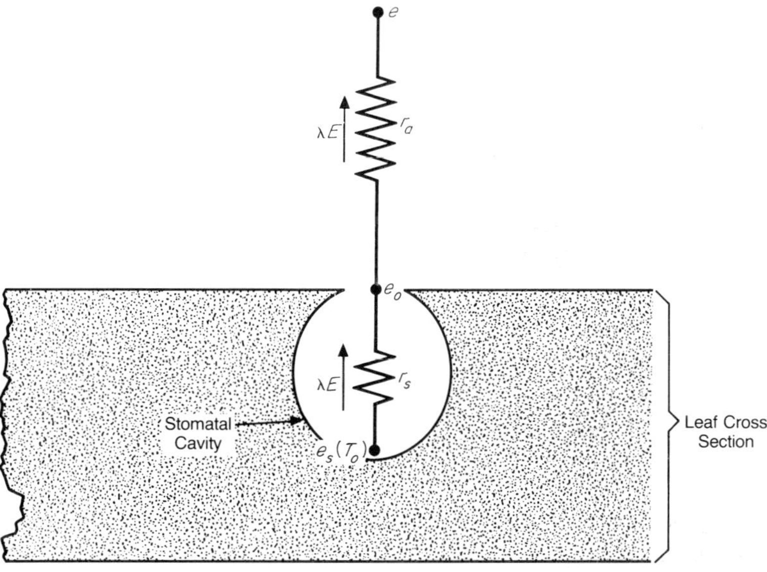

Figure 2.2 Schematic diagram illustrating how the latent heat flux is driven by humidity gradients between the inside of the stomatal cavity ($e_s(T_o)$) and the leaf surface (e_o) and between the leaf surface and the bulk atmosphere (e) against the stomatal and aerodynamic transfer resistances.

Elimination of the surface temperature, surface humidity and sensible heat terms (T_o, e_o, C) from equations 2.1, 2.2, 2.3, 2.4, 2.5 and 2.17 gives the 'Penman–Monteith' equation (Monteith, 1965, 1973):

$$\lambda E = \frac{\Delta H + \rho c_p \delta e / r_a}{\Delta + \gamma(1 + r_s / r_a)} \qquad 2.18$$

Elimination of the surface temperature, surface humidity and sensible heat terms (T_o, q_o, C) from equations 2.1a, 2.2, 2.3, 2.4a, 2.5a, and 2.17a gives the 'Penman–Monteith' equation in terms of specific humidity:

$$\lambda E = \frac{\Delta' H + \rho c_p \delta q / r_a}{\Delta' + c_p(1 + r_s / r_a)/\lambda} \qquad 2.18a$$

The Penman–Monteith form of the combination equation has gained general acceptance by hydrologists, plant physiologists and micrometeorologists for providing a realistic framework for investigating the evaporation rates from vegetation under different environmental conditions. The assumptions involved in the derivation and application of the equation, particularly with respect to the one-dimensional simplification, sometimes described as the 'big leaf' approximation, have been extensively discussed by Thom (1972), Monteith (1981), Jarvis *et al.* (1981) and Stewart (1983). Comparisons with independent measurements and with simulations from more complex multi-layered models have shown that the approximation generally works well (Black *et al.*, 1970; Tan and Black, 1976; Shuttleworth, 1976). Nevertheless, it should be borne in mind that in some circumstances, where the properties of the vegetation canopy are not homogeneous, for example when forest canopies are supporting a melting snow cover, application of the one-dimensional approximation may lead to significant errors. However, for practical purposes, it is probably more important at our current stage of knowledge to determine the magnitude of the errors resulting from the application of one-dimensional models in different circumstances than to try to develop multi-layer models which, because of their excessive data requirements, may never be of use operationally.

2.7 SURFACE TEMPERATURE

2.7.1 Wet surface

The set of equations 2.1, 2.2, 2.3, 2.4 and 2.5 can be solved to give the temperature of a wet surface, (T_o) and surface humidity terms (E, C, e_o):

$$T_o = T + \frac{Hr_a\gamma/\rho c_p - \delta e}{\Delta + \gamma} \qquad 2.19$$

The set of equations 2.1a, 2.2, 2.3, 2.4a and 2.5a can also be solved to give the temperature of a wet surface, in terms of specific humidity (deficit), by elimination of the evaporation, sensible heat and surface humidity terms (E, C, q_o):

$$T_o = T + \frac{Hr_a/\rho\lambda - \delta q}{\Delta' + c_p/\lambda} \qquad 2.19a$$

2.7.2 Frozen surface

For snow or ice at a temperature below the melting point the same method can be applied to calculate the surface temperature if the modified energy balance equation:

$$H = (\lambda + L)E + C \qquad 2.20$$

is introduced, where L is the latent heat of melting. Solution of equations 2.1, 2.2, 2.4, 2.5 and 2.20 then gives:

$$T_o = T + \frac{Hr_a\gamma\lambda/(L + \lambda)\rho c_p - \delta e}{\Delta + \gamma\lambda(L + \lambda)} \qquad 2.21$$

Solution of equations 2.1a, 2.2, 2.4a, 2.5a and 2.20 gives the equivalent relationship in terms of specific humidity (deficit):

$$T_o = T + \frac{Hr_a/\rho(L + \lambda) - \delta q}{\Delta' + c_p/(L + \lambda)} \qquad 2.21a$$

2.7.3 Transpiring surface

For a transpiring surface, the same set of equations, 2.1, 2.2, 2.3, 2.4, 2.5 and 2.17, used for the formulation of the Penman–Monteith equation, can be solved to give the temperature of the vegetative surface by elimination of the evaporation, sensible heat and surface humidity terms, E, C, and e_o:

$$T_o = T + \frac{Hr_a\gamma/\rho c_p - \delta e/(1 + r_s/r_a\rho c_p)}{\Delta(1 + r_s/(r_a\rho c_p)) + \gamma} \qquad 2.22$$

Solution of equations 2.1a, 2.2, 2.3, 2.4a, 2.5a and 2.17a gives the equivalent relationship in terms of specific humidity (deficit):

$$T_o = T + \frac{Hr_a/\rho\lambda - \delta q/(1 + r_s/r_a\rho c_p)}{\Delta'(1 + r_s/r_a\rho c_p) + c_p/\lambda} \qquad 2.22a$$

2.8 EVAPORATION OF INTERCEPTED PRECIPITATION

When vegetation surfaces are covered by precipitation, whether in the liquid or solid state, the surface resistance term in the Penman–Monteith equation (equation 2.18), is effectively 'short-circuited' and the surface resistance, r_s tends to zero. For a wet (liquid precipitation) surface the Penman–Monteith evaporation rate, E_W^{PM}, can be estimated from:

$$\lambda E = \frac{\Delta H + \rho c_p \delta e / r_a}{\Delta + \gamma} \qquad 2.23$$

The equivalent relationship in terms of specific humidity (deficit) is:

$$\lambda E = \frac{\Delta' H + \rho c_p \delta q / r_a}{\Delta' + c_p / \lambda} \qquad 2.23a$$

When the surface is covered by snow or ice (solid precipitation) the latent heat of fusion must be taken into account in the calculation of the evaporation, i.e.:

$$(L+\lambda)E = \frac{\Delta H + \rho c_p \delta e / r_a}{\Delta + \gamma \lambda / (L+\lambda)} \qquad 2.24$$

The equivalent relationship in terms of specific humidity (deficit) is:

$$(L+\lambda)E = \frac{\Delta' H + \rho c_p \delta q / r_a}{\Delta' + c_p / (L + \lambda)} \qquad 2.24a$$

For aerodynamically rough vegetation, such as forests ($r_a \approx$ 3.5 s m^{-1}) the aerodynamic term in the evaporation equation $(\rho c_p \delta q / r_a)/(\Delta' + c_p / \lambda)$ is often much larger than the radiation term $(\Delta' H)/(\Delta' + c_p / \lambda)$ and it is the supply of advected energy, via the cooling of the air mass, rather than the supply of radiative energy that is able to support the high rates of evaporation that are commonly observed from wet forests (Figure 2.3). Indeed high evaporation rates are possible even at night in conditions of low or zero net radiation. The efficiency of forests in making use of this source of energy is so good that in high rainfall upland areas the latent heat requirement for the evaporation of intercepted water may, even on a long-term (annual) basis, exceed the total net radiant energy (Calder, 1979a; Calder and Newson, 1980). For shorter vegetation such as grass ($r_a \approx 50$ s m^{-1}) the

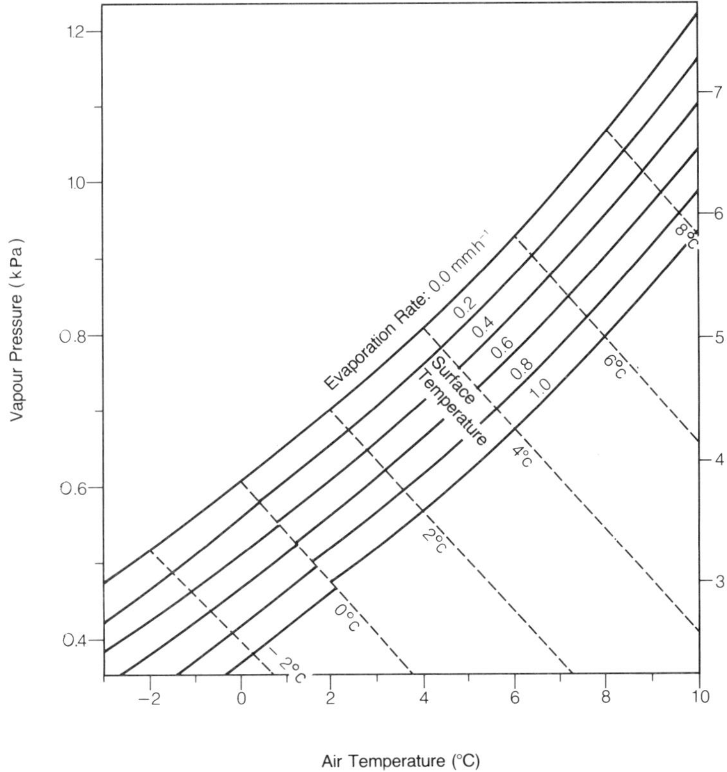

Figure 2.3 Evaporation rates of intercepted precipitation and surface temperatures calculated from solution of the combination equation (equations 2.23a, 2.24a, 2.19a, 2.21a), from forests ($r_a \approx$ 3.5 sm^{-1}) as a function of atmospheric humidity and air temperature when the net radiation is zero. (The effects of surface temperature 'clamping' at the freezing point during snow melt or re-freezing conditions has not been taken into account.)

aerodynamic term is usually smaller than the radiative term and, as expected from theory, observations confirm that the evaporation rates of intercepted water from short vegetation are closely related to the radiant energy supply and are similar to the equilibrium rate.

2.9 TRANSPORT THEORY

2.9.1 Eddy diffusion theory

The most widely used model for the transport of entities, principally momentum, heat and water vapour, between the surface and the atmosphere involves the assumptions that:

(1) The flux is one-dimensional, normal to the surface.
(2) The flux vector is proportional to the negative of the gradient vector.
(3) The gradient vector is inversely proportional to the distance from an arbitrary surface displaced from the real surface by a distance d.

Conventionally these assumptions are formally expressed by the following equations:

$$\frac{du}{dz} = \frac{u_*}{k(z-d)} \qquad 2.25$$

$$\frac{dT}{dz} = \frac{T_*}{k(z-d_T)} \qquad 2.26$$

$$\frac{dq}{dz} = \frac{-q_*}{k(z-d_q)} \qquad 2.27$$

where k is a dimensionless constant, von Karman's constant, the value of which has been determined experimentally as 0.41; d, d_T and d_q are termed the zero plane displacements for momentum, temperature and humidity respectively and are often assumed equal (although for tall vegetation this may not be valid); U_*, T_* and q_* are wind, temperature and humidity scales related to the fluxes of momentum (M), heat and water vapour via:

$$M = \rho u_*^2 \qquad 2.28$$

$$C = \rho c_p U_* T_* \qquad 2.29$$

$$E = -\rho u_* q_* \qquad 2.30$$

Equations 2.25 to 2.27 can be integrated to give:

$$u(z) = \frac{u_*}{k} \ln((z-d)/z_o) \qquad 2.31$$

$$T(z) - T(0) = \frac{T_*}{k} \ln((z-d_T)/z_T) \qquad 2.32$$

$$q(z) - q(0) = \frac{q_*}{k} \ln((z-d_q)/z_q) \qquad 2.33$$

where the constants of integration, z_o, z_T, z_q, are termed the roughness lengths for momentum, temperature and humidity respectively (and, like the roughness lengths for the different entities, are often assumed equal).

These equations can also be used to calculate the transfer coefficients for the various entities. By definition the transfer coefficient for momentum (K_M) at height z is given by:

$$K_M(z) = \frac{M dz}{d(\rho u)} \qquad 2.34$$

which, on substituting equations 2.25 and 2.28, gives:

$$K_M(z) = ku_*(z-d) \qquad\qquad 2.35$$

which implies that the transfer coefficient increases linearly with height above the zero plane displacement. (Similar equations are obtained for the other entities.)

The aerodynamic resistance to the transport of momentum, r_{aM}, from the level of the momentum sink, $d + z_o$ (where $u = 0$) to a height z can also be calculated from these equations. Again by definition:

$$r_{aM} = \frac{\rho u(z) - \rho u(d + z_o)}{M} \qquad\qquad 2.36$$

which on substituting equations 2.28 and 2.31 and noting that $\rho u(d + z_o) = 0$ gives the useful relation:

$$r_{aM} = \frac{u(z)}{u_*^2}$$

$$= \frac{\ln(z-d)/z_o}{ku_*^2}$$

$$= \frac{(\ln(z-d)/z_o)^2}{k^2 u(z)} \qquad\qquad 2.37$$

which implies, if the roughness length, z_o, is independent of windspeed, an inverse relationship with the mean windspeed.

The term u_*, sometimes called the 'friction velocity', can be interpreted physically as an index of the rate of rotation of the frictionally driven eddies in the airflow above the surface (Thom, 1975). In conditions of positive thermal buoyancy the frictionally driven eddy transport mechanism is enhanced and various empirical stability functions have been proposed to take this into account (Thom, 1975).

For short vegetation and other relatively smooth surfaces the model predicts logarithmic profiles of humidity, temperature and momentum which agree well with observations. When the surface is composed of taller elements, such as trees, or when the local topography is rough the model is not altogether satisfactory; countergradient fluxes are sometimes observed and other model predictions, such as the windspeed dependence of the transport coefficient, are not in agreement with observation (see section 6.7). A number of possible explanations for these deficiencies has been suggested. The height of the planetary boundary layer (PBL) may be insufficient for the full development of a logarithmic profile when trees of, say, 30 m high are generating eddies which may be only an order of magnitude or so less than the height of the PBL. Another, possibly related, explanation involving gust penetration is discussed below.

2.9.2 Gust penetration

An additional transport mechanism, which has been observed or inferred to exist in a number of forests (Denmead, 1984; Crowther and Hutchings, 1984; Calder and Wright, 1986), operates by the 'piston' displacement of air, water vapour and heat from within the canopy airspace by large gusts or eddies which, although only present for short periods of time, are very efficient as transporters of these entities. Observations using smoke as a tracer (from smoke bombs ignited on the ground surface) have shown that the process occurs in unstable conditions when thermal plumes are likely to occur (Oliver, 1975) and also in neutral or stable conditions when forests are wet (Figures 2.4 and 2.5). The scale lengths appear to be of the order of tens to hundreds of metres. To what extent these gusts are associated with thermal- or humidity-driven plumes, the local topography or other factors is still unknown; a satisfactory theory of the mechanisms involved has yet to be developed.

2.10 EVAPORATION–ATMOSPHERE INTERACTION

2.10.1 Processes

Changes in either atmospheric demand or the water availability at the ground surface will alter sensible and latent heat fluxes and the temperature and latent heat profiles near the surface. The extent to which the bulk properties of the planetary boundary layer (PBL) are altered by these changes is a study area of major importance but one which, lying between the near surface 'logarithmic profile' atmospheric model and the larger scale 'circulation' models, is at present treated inadequately. Great advances have been made recently by McNaughton (1976), de Bruin (1983) and McNaughton and Spriggs (1986) although their models are not applicable in all climatic conditions and not, unfortunately, in the frontal conditions that are common in the UK, especially when forests are wet.

A knowledge of the processes involved is important because greater insight will be of value in a number of applications including: (a) general circulation modelling; (b) assessing the effects of forest clearances on 'desertification' in arid areas; and (c) estimating the—possibly reducing—interception loss rates that would be expected from forests of increasing areal extent.

The relevance or importance of the feedback between evaporation flux and the atmosphere is not in doubt but there is considerable uncertainty over the scale and extent of the process.

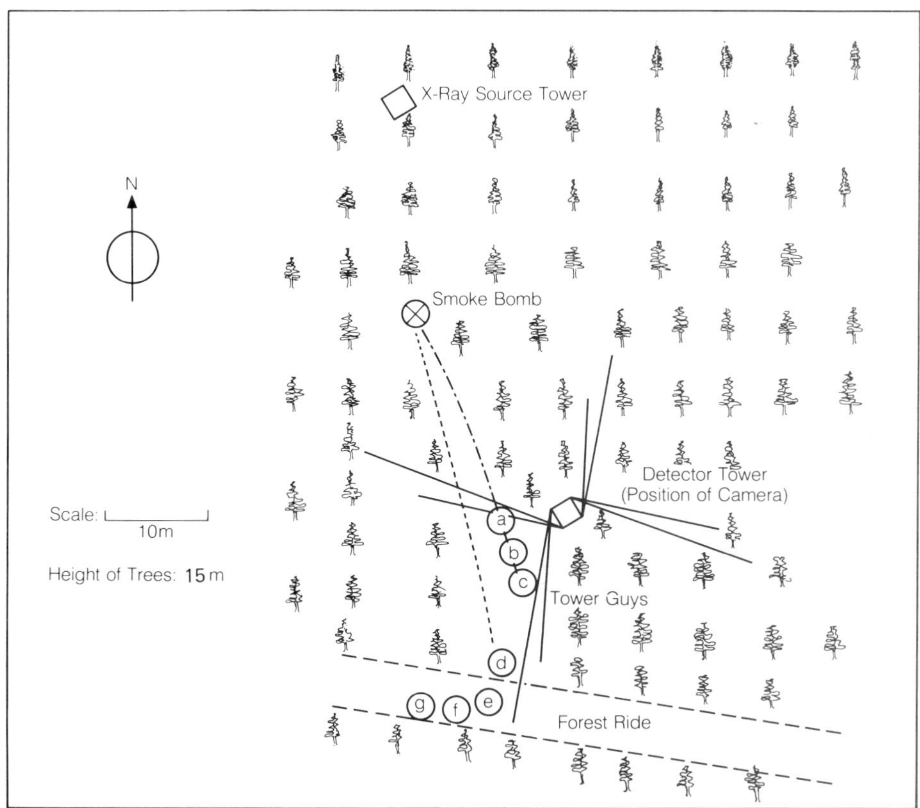

Figure 2.4 Schematic diagram which, in conjunction with Figure 2.5(a–g), shows the movement and coherent nature of smoke plumes formed as a result of the ignition of a smoke bomb on the ground surface at the γ-ray site in the Queens Forest, Aviemore at 15.30 on 5 June 1985 in conditions when the forest canopy was wet and the atmosphere was in a condition of neutral stability (air temperature = 6.5 °C, net radiation = 28 W m^{-2}, specific humidity deficit = 5 g kg^{-1}, wind speed = 2.3 m s^{-1}).

Morton (1985) has proposed that the mechanism operates on length scales which are of the order of hundreds of metres, whereas Calder (1987b) has suggested that the scale is more likely to be of the order of kilometres or tens of kilometres.

It is important to take into account the two distinct (but often confused) mechanisms which are involved when a change in surface characteristics brings about alterations in the atmospheric conditions above the surface.

2.10.2 Mechanism 1: 'Logarithmic profile'

Firstly, if we consider the 'logarithmic profile' model of evaporation (see e.g. Thom, 1975) the water availability at the surface (treated in the Penman–Monteith equation as a surface resistance

Figure 2.5 (a–g) Smoke bombs in use in the Queens Forest, near Aviemore, to investigate the transfer mechanisms in wet conditions. The photographs were taken at 2-minute intervals after the ignition of the bomb.

r_s) may be considered to act to partition net radiation into sensible and latent heat and, depending upon the aerodynamic transfer resistance of the crop, the necessary gradients of temperature and humidity for transporting these fluxes will be imposed at the

Figure 2.5 (*continued*)

surface. The gradients will decay exponentially with increasing altitude until a level is reached, which may be at the top of the planetary boundary layer, where no further decrease occurs.

A change in surface characteristics, either through water availability (r_s) or through aerodynamic roughness (r_a) will therefore alter the gradient relationships at the surface, and

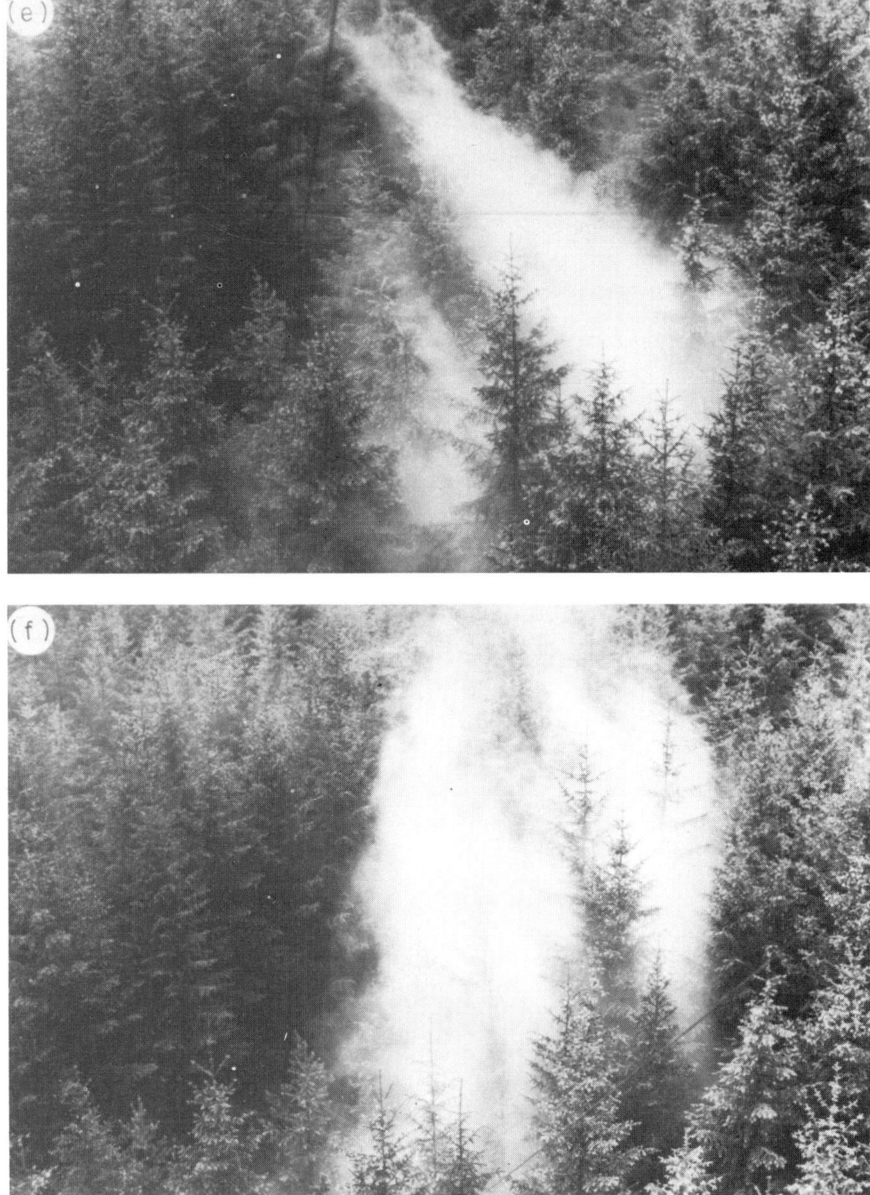

Figure 2.5 (*continued*)

measurable differences in the meteorological conditions immediately above the surface will be apparent. However, these differences will reduce with increasing altitude (see e.g. McNaughton and Jarvis, 1983). When two different vegetation types are adjacent, a region will exist where the gradient relationships will be readjusting.

Figure 2.5 (*continued*)

2.10.3 Mechanism 2: 'closed box'

Secondly, there is the mechanism, described by Morton (1985), in which, in stable conditions, evaporation from the surface is expected to humidify and cool essentially a 'closed box' of air which moves across the surface until saturation is reached and further evaporation will be governed solely by the availability of radiant energy and not advected energy.

Experimentally it may not always be easy to determine which of these mechanisms is at work or which is the more important. There is, however, reason to believe that both mechanisms exist; what is in doubt is the typical length scales over which they operate. Conventionally it is thought that where a change from grassland to forest occurs, profiles will readjust after distances of the order of 100–200 m (see e.g. Gash, 1985). The length scale for the second mechanism is less well understood; the evidence which is reviewed by Calder (1987b) suggests that it is at least a factor of ten greater than that proposed by the Morton model with a 'closed box' depth of 25 m, and operates over distance scales of the order of kilometres rather than hundreds of metres.

Essentially the evidence is that:

(a)   High rates of evaporation, exceeding the net radiation, are found from wet forests in the UK in areas of both high and low relief, that is, at both Plynlimon and Thetford.

Typically the dimensions of UK forests are of the order of 10–20 km. This suggests that the scale length is of the same order or greater than the dimensions of these forests.

(b)   At Plynlimon annual evaporation from the forest can exceed the net radiation equivalent of evaporation (which gives further confirmation of (a).

(c)   Harding (personal communication) has shown from comparison of data from automatic weather stations situated both above forest and grassland that vapour pressure deficits in wet conditions are indeed less above forests but the lessening is due to a reduction of air temperature, not an increase in absolute humidity. This result suggests that the differences are mainly due to mechanism 1; mechanism 2 would indicate a lowering of temperature but an increase in atmospheric humidity. However, Harding, making use of limited wet condition data from a 'two-site' experiment at Thetford, found some evidence for mechanism 2: a reduction in temperature of 0.2 °C was found between a site 0.5 km downwind from the edge of the forest as compared with a site 2 km downwind (but again no measurable differences in humidity were found). Using this data to calibrate Morton's 'closed box' model he found that the height of the box was 200 m (about ten times that suggested by Morton).

(d)   The theoretical justification for the assumption of a 'closed box' height of 25 m is tenuous. Morton suggested (following McNaughton and Jarvis, 1983), that in conditions when forests are wet, temperature inversions will occur and evaporation will take place into a restricted mixed layer depth. In mountainous topography he concedes that mixing may occur to much greater heights. However, observations at Plynlimon and Thetford (Stewart, personal communication) indicate that the strong inversions, as described by McNaughton and Jarvis (1983), are not common in wet conditions; generally in wet conditions the PBL is close to neutral stability, probably because there is sufficient wind, especially over an aerodynamically rough forest, to generate turbulence throughout the PBL and allow 'leakage' of water vapour from the PBL to higher levels in the atmosphere. The McNaughton and Jarvis representation of an inversion in stable conditions (their Figure 5) may more adequately describe a nocturnal inversion, induced by the conditions (untypical of those when forests are wet) of low windspeeds, clear skies and negative net radiation.

An understanding of the feedback processes between evaporation from surface vegetation and atmospheric conditions represents one of the greatest challenges and study areas for environmental science; its solution will only be achieved through the combined efforts of workers in the related disciplines of meteorology, hydrology, plant physiology and soil physics.

CHAPTER 3

The Interception Process

3.1 INTRODUCTION

In upland areas the evaporation of water from wet vegetative surfaces forms a major part of the total evaporation; from forests it is the major part and may typically be about twice the component due to transpiration. The importance attached to understanding the processes involved cannot be overstated and no apologies are given for the detailed discussion given to this subject.

Although perhaps more studies have been carried out on interception than any other single subject area in hydrology, the increase in understanding has not been commensurate and the mechanisms remain poorly understood. Most of the early studies attempted to relate interception losses to rainfall amounts through simple empirical relationships. The pioneering work of Rutter and colleagues (Rutter *et al.*, 1971, 1975) in the early 1970s broke new ground by considering the water balance of stored precipitation on the canopy and relating it to rates of precipitation, drainage and evaporation. Only recently has this method been reconsidered as a result of a rather different stochastic description of canopy wetting.

The different theoretical treatments of interception are given in this chapter, and later chapters describe the experimental studies which were carried out to determine the interception model parameters for different vegetation types in both rainfall and snowfall conditions.

3.2 THE RUTTER INTERCEPTION MODEL

In the early 1970s Rutter and colleagues proposed a revolutionary process-oriented interception model which explicitly and quantitatively described the water balance of rain intercepted by the leaves and stems of vegetation. Inputs to canopy storage were assumed equal to the precipitation rate minus a constant 'free throughfall fraction' (p). Drainage was uniquely related to canopy storage via two drainage parameters (b and k), and evaporation was assumed proportional to canopy storage for storage less than the canopy capacity (S) value.

The model can be summarized in terms of the continuity equation:

$$dc/dt = Q - k\,(\exp(bc) - 1) \qquad 3.1$$

where:

$$Q = (1-p)R - E_W^{PM}c/S \quad \text{for } c/S \qquad 3.2$$

$$Q = (1-p)R - E_W^{PM}, \quad \text{for } c \geq S \qquad 3.3$$

where:

c = canopy storage (mm)
R = precipitation rate (mm min^{-1})
b = drainage parameter (mm^{-1})
k = drainage parameter (mm min^{-1})
p = free throughfall fraction (dimensionless)
S = canopy storage capacity (mm)
Q = rate of precipitation plus evaporation (mm min^{-1})
E_W^{PM} = Penman–Monteith evaporation rate in wet conditions calculated from equation 2.18a (mm min^{-1})

3.3 THE STOCHASTIC INTERCEPTION MODEL

3.3.1 Background

A stochastic model of rainfall interception was proposed (Calder, 1987a) to account for recent observations, from both laboratory studies using rainfall simulators (Aston, 1979) and field studies (Calder *et al.*, 1987), which were irreconcilable with the conventional Rutter type methods. Essentially these observations are:

(a) That for the same canopy storage, drainage rates are higher in the early, wetting-up stages of the storm as compared with the later, drying-out phase.

(b) That maximum canopy storage is approached in a 'gradual' manner and is only obtained after application of a considerably greater depth of rainfall than the capacity value.

Aston (1979) proposed that maximum canopy saturation was approached in an exponential manner as cumulative rainfall increased, i.e.:

$$c = c_{max}[1 - \exp(-k'P/c_{max})] \qquad 3.4$$

where:

$$c_{max} = \text{maximum storage depth} \qquad \text{(mm)}$$
$$k' = \text{drainage parameter} \qquad \text{(dimensionless)}$$
$$P = \text{cumulative depth of rainfall} \qquad \text{(mm)}$$

Herwitz (1985) suggested, by way of explanation, that 'the filling of a plant's interception storage appears asymptotic because of the gradual accumulation of rainsplash on the sheltered undersides of leaves and branches.'

An additional, or alternative, explanation is that the manner in which the individual elements of the surface of the tree are struck and wetted by individual raindrops is governed by the laws of probability.

3.3.2 Stochastic hypothesis

If the surface area of a tree is considered to be composed of elemental areas, each of which has the same probability of being struck by raindrops, and the mean number of raindrop strikes per element during a rainstorm (without evaporation) is m, the Poisson probabilities P_o, P_1, P_2, $..P_r$, .. of elements being struck by 0, 1, 2, $..r,.$ drops is then given by:

$$e^{-m}, \frac{me^{-m}}{1!}, \frac{m^2e^{-m}}{2!}, \cdots \frac{m^re^{-m}}{r!} \qquad 3.5$$

The cumulative probability of elements being struck by 0, 1, 2, $..r,.$ drops is given by:

$$\sum_{x=0}^{r} P_x = e^{-m} + \frac{me^{-m}}{1!} + \frac{m^2e^{-m}}{2!} + ..\frac{m^re^{-m}}{r!} \qquad 3.6$$

and the probability of elements being struck by more than r drops is therefore:

$$1 - \sum_{x=0}^{r} P_x$$

If the elemental areas can retain on average q drops, elements receiving more than r strikes, where r is the integer equal to the truncated value of q, will shed the excess by drainage. The mean number of drops retained per element (n) is then given by:

$$n = (1 - \sum_{x=0}^{r} P_x)q + 1\frac{me^{-m}}{1!} + 2\frac{m^2e^{-m}}{2!} + ..r\frac{m^re^{-m}}{r!}$$

where the first term represents the contributions from the fully wet elements and the succeeding terms are the contributions from the partially wet elements. On rearrangement this gives:

$$n = q + (0-q)e^{-m} + (1-q)\frac{me^{-m}}{1!} + (2-q)\frac{m^2e^{-m}}{2!} +$$

$$..(r-q)\frac{m^re^{-m}}{r!} \qquad 3.7$$

The manner in which n increases with increasing m is shown in Figure 3.1 for selected values of q. The same relationship normalized by the mean maximum drop retention value q is shown in Figure 3.2, again for a selection of q values.

Differentiating equation 3.7 with respect of m yields:

$$\frac{dn}{dm} = qe^{-m} + \frac{(1-q)}{1!}(1-m)e^{-m} + \frac{(2-q)}{2!}(2m-m^2)e^{-m} +$$

$$..\frac{(r-q)}{r!}(rm^{(r-1)}-m^r)e^{-m} \qquad 3.8$$

which is used below.

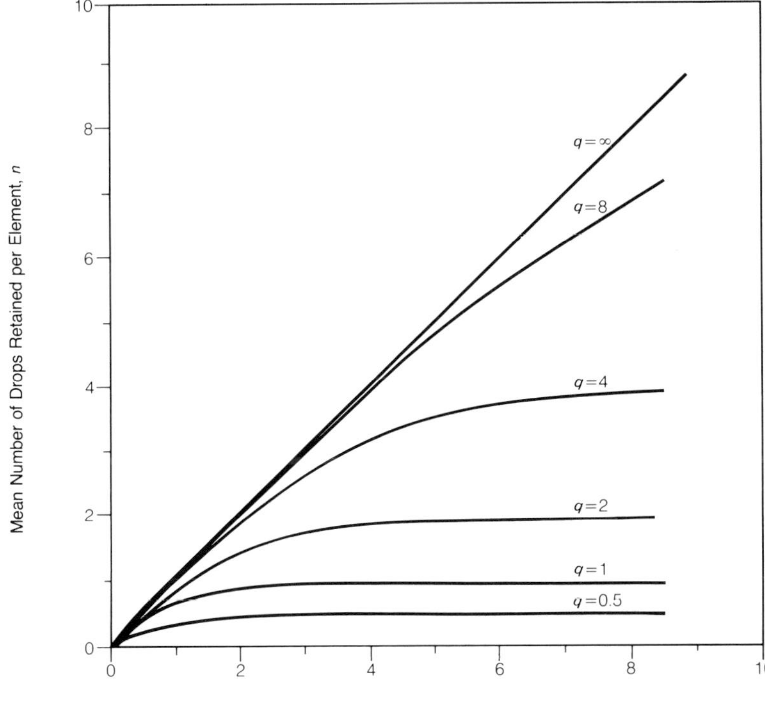

Figure 3.1 The relationship between n, the mean number of drops retained per element, and m, the mean number of raindrop strikes per element for selected values of q, the maximum retention number of drops per element. (Reproduced by permission from Calder, 1987a; Copyright Elsevier Science Publishers B.V.)

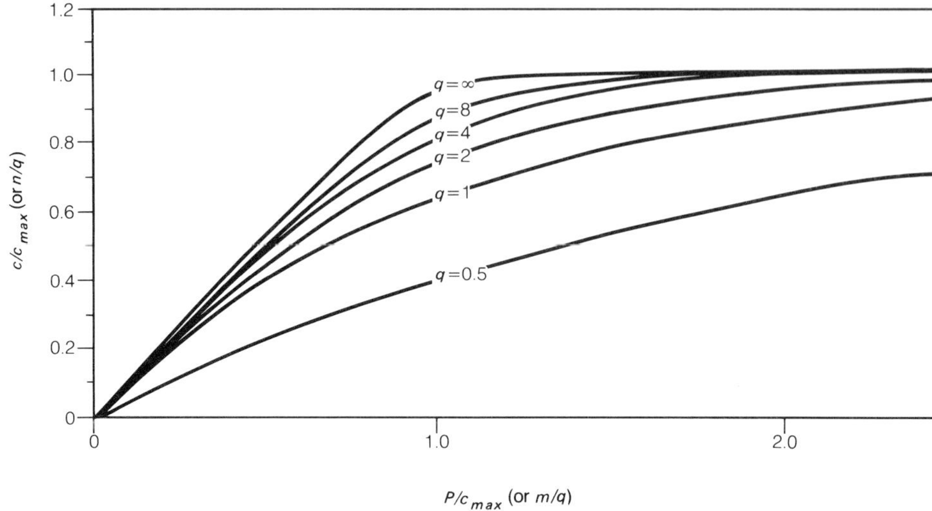

Figure 3.2 The dimensionless 'wetting' relationship between c/c_{max} and P/c_{max} (or identically, n/q and m/q) for different values of q. (Reproduced by permission from Calder, 1987a; Copyright Elsevier Science Publishers B.V.)

3.3.3 Whole canopy accounting model

The development of a whole canopy accounting model requires re-scaling from the element to the canopy scale and the incorporation of evaporation.

Scaling involves the introduction of a further canopy parameter, L, the number of elemental surface areas per unit ground area, and an input variable, v, the mean volume of raindrops. The depth of storage on the canopy on a ground area basis, c, is then given by:

$$c = nvL \qquad\qquad 3.9$$

The maximum depth of canopy storage is given by:

$$c_{max} = qvL \qquad\qquad 3.10$$

and the mean number of raindrop strikes per element is related to the rainfall depth, P, by:

$$m = P/(vL) \qquad\qquad 3.11$$

Evaporation of a depth of intercepted water, I, will result in an equivalent reduction in canopy storage, c, i.e.:

$$\frac{dc}{dI} = -1 \qquad\qquad 3.12$$

The rate of change of n with respect to evaporation is then given by (equation 3.12 and on differentiating equation 3.9):

$$\frac{\mathrm{d}n}{\mathrm{d}I} = \frac{\mathrm{d}c}{\mathrm{d}I} \times \frac{\mathrm{d}n}{\mathrm{d}c} = \frac{-1}{vL} \qquad 3.13$$

A rational method for incorporating evaporation explicitly into the model is now available, if m is redefined as a state variable which represents the equivalent mean number of raindrop strikes per element. Its value will reduce as a result of evaporation and will increase during rainfall. If a finite differencing technique is adopted, the incremental change in m (Δm) can be calculated as a result of depth applications ΔP of rainfall and ΔI of evaporation:

$$\Delta m = \frac{\mathrm{d}m}{\mathrm{d}P}\Delta P + \frac{\mathrm{d}m}{\mathrm{d}I}\Delta I$$

$$= \frac{\mathrm{d}m}{\mathrm{d}P}\Delta P + \frac{\mathrm{d}n}{\mathrm{d}I}\Delta x \frac{\mathrm{d}m}{\mathrm{d}n}\Delta I$$

which, from equations 3.8 and 3.13, and on differentiating equation 3.11:

$$\Delta m = \frac{1}{vL}\Delta P - \frac{1}{vL}(1/[+qe^{-m} + \frac{(1-q)}{1!}(1-m)e^{-m} +$$

$$\frac{(2-q)}{2!}(2m-m^2)e^{-m} + .. \frac{(r-q)}{r!}(rm^{(r-1)}-m^r)e^{-m}])\,\Delta I \qquad 3.14$$

Thus with an accounting procedure to calculate m during a storm (equation 3.14) canopy storage can be calculated from equations 3.7 and 3.9 and net-rainfall can be obtained by solving the conventional water balance equation.

Unlike the pioneering interception models developed by Rutter and colleagues (Rutter *et al.*, 1971) the stochastic model does not calculate drainage from the canopy in terms of a unique relationship with canopy storage; rather it predicts that the proportion of rain shed from the canopy is uniquely related to the proportional canopy storage, c/c_{max} (Figure 3.3).

An example of the operation of the model is the special case where the maximum number of raindrops retained (q) is equal to or less than unity. Equations 3.9 and 3.7 then reduce to:

$$c = qvL(1-e^{-m}) \qquad 3.15$$

or, substituting equations 3.10 and 3.11:

$$c = c_{max}[1\text{-}\exp(qP/c_{max})] \qquad 3.16$$

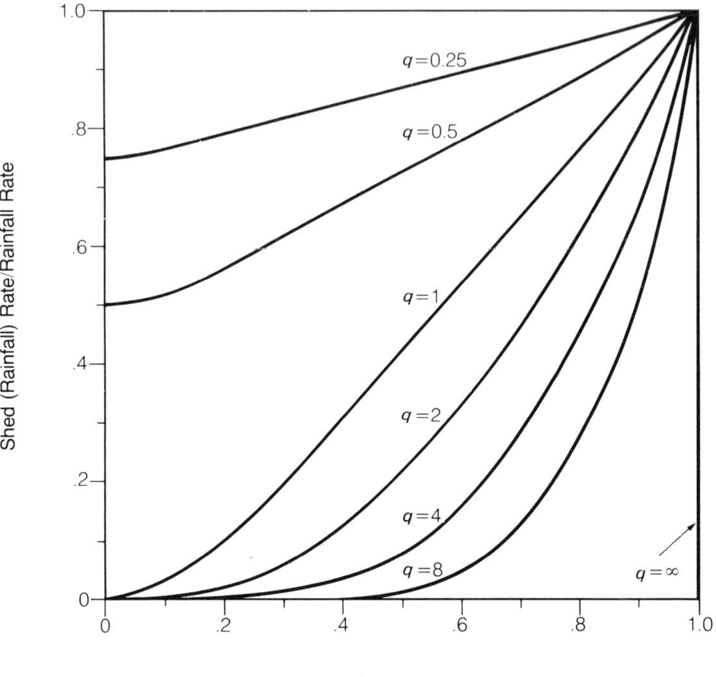

Figure 3.3 The dimensionless relationship between the instantaneous proportion of rain shed from the canopy and canopy storage. (Reproduced by permission from Calder, 1987a; Copyright Elsevier Science Publishers B.V.)

which is identical to the Aston (1979) formulation (equation 3.4), with Aston's k' value equal to q, the mean number of raindrops retained per element. With this interpretation of Aston's k' values, the mean number of raindrops held per surface element ranged between 0.16 (*E. cinerea*) and 0.77 (*E. maculata*) for *Eucalyptus* species and was unity (or, with the stochastic interpretation, possibly greater than unity) for *Pinus radiata* for drops of 2.5 mm diameter.

Drops as large as 2.5 mm diameter would, at least for temperate climates, represent the very upper part of the spectrum of drop sizes occurring naturally during rainstorms. Drop volume is an input variable to the stochastic model and the effects of differing raindrop sizes on the wetting function can be inferred from Figure 3.2. For example, a variation in drop radius of a factor of 3 (and hence of drop volume of 27) could result in a difference in the wetting-up responses equivalent to q varying from 1 to 27.

Under natural conditions, therefore, if raindrop sizes are smaller than those used by Aston in his simulator, canopies would require the application of very much less rainfall to produce the same degree of wetting.

For parsimony of model parameters, the model, as proposed, does not take into account explicitly either the different and reduced probabilities that shaded elements of the canopy will have for receiving raindrop strikes, or the routing of shed water from the canopy. In the first case, the principal effect of a range of strike probabilities will be to increase the variability in the drop retention distribution within the elemental areas; this effect may be approximately and implicitly accounted for by reducing the q parameter value. For the second, the introduction of a linear routing reservoir may suffice. Furthermore, for open canopies the introduction of a 'free throughfall parameter' as used and described by Rutter *et al.* (1971) may be necessary (although it is interesting to note that if q values are less than or equal to unity, the Aston formulation, the effects of a non-zero p value, are fully represented simply by a lowering of the q value.

3.4 THE SIMPLIFIED 'DAILY' STOCHASTIC MODEL

For a storm in which no evaporation occurs during precipitation and the total precipitation is P, the total interception loss will be equal to the storage remaining on the canopy at the end of the storm and can be calculated from equation 3.16 in terms of P and the two model parameters c_{max} and q.

However, in reality, and particularly in regions like the uplands of the UK, where long-duration, low-intensity rainstorms are common, the evaporation occurring during precipitation is a significant proportion of the total interception. One approach to estimating the evaporation during precipitation, if meteorological data are unavailable, is to assume, following the reasoning of Calder and Newson (1979) and Gash (1979), that:

(a) the rate of evaporation in wet conditions is a conservative quantity both in space and time, and that
(b) the total evaporation during precipitation can be estimated from the product of $\bar{E}$, the mean evaporation rate, and the duration of rainfall, where the duration of rainfall is itself estimated from the total precipitation (P) divided by the mean rainfall intensity, $\bar{R}$.

With this simplifying assumption it is then possible to calculate, using the whole canopy accounting model (equations 3.7, 3.9 and 3.14), the total storm interception loss I in terms of q, c_{max}, P and the ratio of the mean rates of evaporation and precipitation, $\bar{E}/\bar{R}$.

When $\bar{E}/\bar{R} = 0$, then $I = c$ and the relationship between I and c_{max} and P reduces to that shown in Figure 3.2. The equivalent relationship for $\bar{E}/\bar{R} = 0.125$ is shown in Figure 3.4.

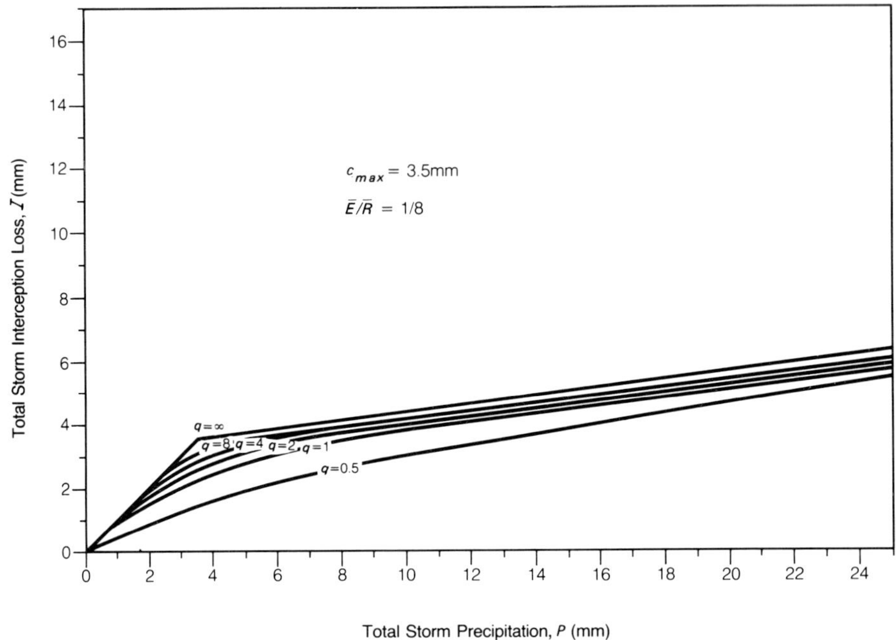

Figure 3.4 Predictions of daily interception loss as a function of daily precipitation for $c_{max} = 3.5$, $\bar{E}/\bar{R} = 0.125$, and a range of values of q, simplified daily stochastic model.

CHAPTER 4

Further Transpiration Studies in the Uplands

4.1 INTRODUCTION

Although interception is responsible for the major differences in evaporation between the different upland vegetation types, and, commensurately more studies have been carried out to measure and model this process, differences in transpiration are also important. It could be argued that during the dry and possibly drought periods when the availability of water for supply purposes or for maintaining minimum flows in rivers is most crucial, transpiration differences are equally important. A number of small-scale studies have recently been carried out in the British uplands, both in the wetter western and drier eastern coastal regions; together these studies are beginning to give a fairly consistent and comprehensive picture of the different transpirational characteristics.

4.2 THE STOCKS HEATHER AND GRASS LYSIMETERS

Frank Law's important and controversial results from his forest lysimeter and catchment experiments at Stocks Reservoir provided the impetus for many of the subsequent evaporation studies in the British uplands. A number of other, less well known, meteorological and hydrological studies were also carried out at Stocks and these were located within the confines of a second climatological station, site 10, which was installed in 1955 on an exposed promontory overlooking the reservoir, Figure 4.1. The average wind speed at this site was twice that recorded at the original climate station, situated near to the forest lysimeter. The instrumentation included experimental raingauges and two heather and two grass weighing lysimeters (Figure 4.2), and two grass 'transpiration gauges' (essentially drainage lysimeters which were

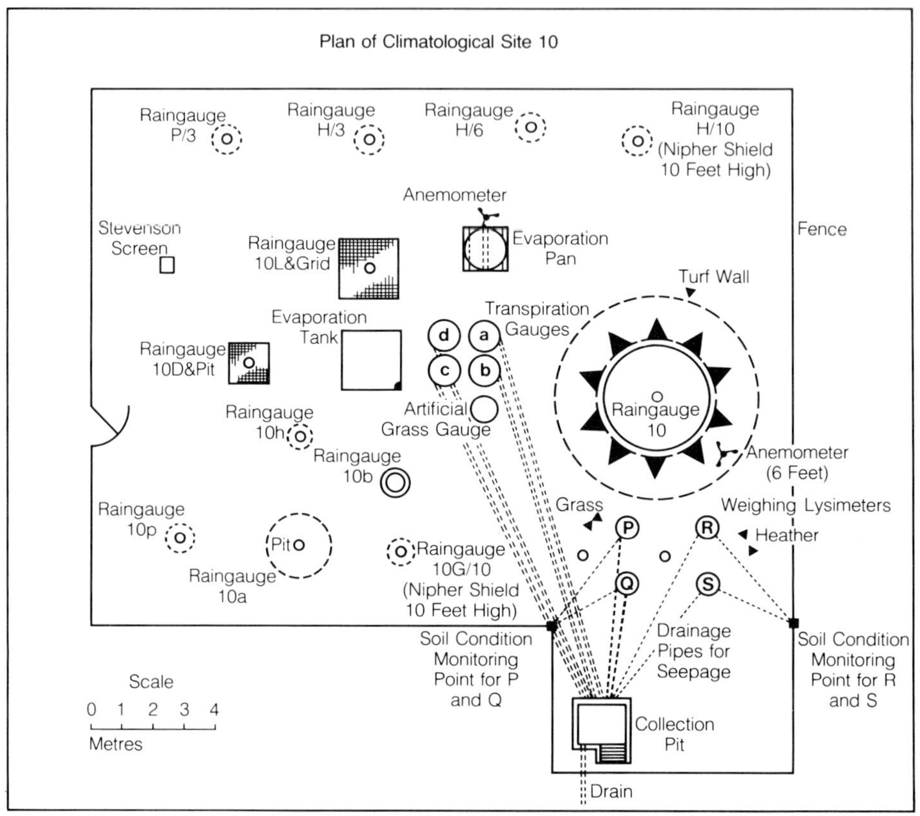

Figure 4.1 Layout of hydrological experiments at climatological site 10, Stocks Reservoir.

kept at field capacity by irrigation). Unfortunately, the analysis of these experiments was not completed by Law and no results were published. The original field sheets relating to these innovative experiments and to the forest lysimeter over an extensive time period were later discovered on a visit to the original site in 1980 by Dr Walsh (of North West Water), Dr Newson and the author (both of the Institute of Hydrology), and these have formed the basis for a number of studies (Calder *et al.*, 1982b; Walsh and Walker, 1985; see also section 1.2).

The results from the heather and grass lysimeters were not easy to interpret because of deficiencies in representivity and because of the experimental errors inherent in the method. In particular, the 0.25 m² surface area heather lysimeters were set within the short mown grass of a meteorological site (Figure 4.2) and might be expected to suffer from aerodynamic edge effects. Errors in the drainage measurement caused by freezing of outlets and leaks were at times suspected from the results but could not be confirmed because of the long time interval since the measurements were taken and the lack of first-hand knowledge

Figure 4.2 Heather and grass lysimeters at Stocks Reservoir.

of the experiments. Furthermore, the spatial variability and the difficulty in measuring the water equivalent of snow precipitation invalidated the measurement of evaporation for periods with significant snow precipitation and these have been excluded from the analysis.

However, although the quality of the data may not be sufficient to provide a conclusive result, data on evaporation from different vegetation types in the uplands are very limited and the Stocks results, encompassing 4 years of useful data between 1964 and 1968 from the heather and grass lysimeters, represent a unique data set which can at least be used to provide a basis for comparison with later studies.

The water balances of the heather and grassland lysimeters are given in Table 4.1 and shown in Figure 4.3. The annual total water use is similar and, under the climatic regime at Stocks, where the average annual rainfall is 1650 mm, the result is consistent with the daily evaporation model that is described in section 9.3.

A simple modelling exercise was carried out on the data from the heather lysimeters in which two parameters, the interception ratio, α, and the transpiration ratio, β, (the ratio of actual transpiration to the Penman (1948) E_T potential transpiration estimate for grass), were optimised. The objective function that was minimised was composed of the sum of squares of the differences between predicted and observed daily drainage from the lysimeters. The results of this procedure gave α values ranging

Table 4.1 Summary of data from the Stocks grass (tanks P and Q) and heather (tanks R and S) drainage lysimeters

Field capacity return period (day no./year)		Length (days)	Rainfall (mm)	Evaporation (mm)				Mean evaporation		Penman E_T
				Grass		Heather		Grass (mm)	Heather (mm)	(mm)
				P	Q	R	S			
33–56	65	23	16.9	4.3	5.6	9.5	11.1	4.9	10.3	15.2
56–76	65	20	29.8	31.7	31.9	26.2	29.4	31.8	27.8	14.2
76–89	65	13	57.1	31.7	31.9	27.2	29.4	31.8	28.3	15.2
89–114	65	25	102.4	33.3	31.9	12.3	14.9	32.6	13.6	47.8
114–146	65	32	91.6	45.1	44.0	44.0	50.2	44.6	47.1	81.0
146–167	65	21	33.5	24.5	24.6	30.1	31.4	24.5	30.8	60.1
167–200	65	33	116.8	57.1	62.5	89.3	86.5	59.8	87.9	93.4
200–225	65	25	148.5	54.6	54.7	65.0	62.5	54.7	63.8	59.3
225–266	65	41	183.8	64.6	69.1	73.5	83.2	66.8	78.3	64.6
266–295	65	29	93.3	29.7	31.8	38.1	28.8	30.8	33.5	26.8
295–326	65	31	117.3	19.2	14.5	41.3	24.5	16.8	32.9	16.1
TOTAL		293	990.8	395.8	402.5	456.5	451.9	399.1	454.3	493.7

8–79	67	71	310.9	35.2	29.8	5.7	5.5	32.5	5.6	65.3
79–99	67	20	57.7	24.3	24.3	20.6	19.0	24.3	19.8	40.6
99–160	67	61	211.2	73.2	71.8	68.7	65.4	72.5	67.0	132.6
160–186	67	26	96.8	44.3	50.0	66.9	68.3	47.2	67.6	75.6
186–209	67	23	69.5	38.7	39.6	61.8	62.2	39.2	62.0	60.9
209–238	67	29	266.8	63.7	73.5	51.1	42.9	68.6	47.0	56.2
238–258	67	20	138.2	33.6	33.5	45.4	41.4	33.5	43.4	34.5
258–365	67	107	837.0	137.5	131.8	175.1	134.8	134.6	154.9	60.9
TOTAL		357	1988.1	450.5	454.3	495.3	439.5	452.4	467.3	526.6
0–54	68	54	258.3	10.7	6.2	31.5	15.9	8.5	23.7	19.6
54–71	68	17	35.7	32.4	32.2	32.3	31.9	32.3	32.1	16.9
71–104	68	33	293.4	47.0	46.3	77.0	79.5	46.6	78.2	47.3
104–154	68	50	190.0	60.1	59.9	54.7	50.7	60.0	52.7	116.2
154–195	68	41	185.0	83.2	83.7	101.1	90.5	83.5	95.8	110.9
TOTAL		195	962.4	233.4	228.3	296.6	268.5	230.9	282.5	310.9
GRAND TOTAL		845	3941.3	1079.7	1085.1	1248.4	1159.9	1082.4	1204.1	1331.2

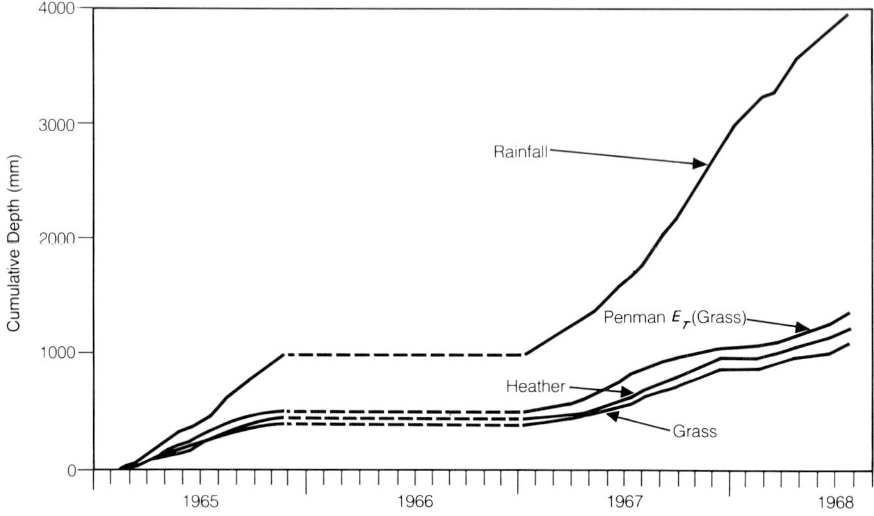

Figure 4.3 Rainfall, Penman's E_T (grass) and measured evaporation from the grass and heather drainage lysimeters, accumulated over time periods when the lysimeters returned to field capacity for most of years 1965 and 1967 and the early part of 1968.

from 0.13 to 0.18, with a mean of 0.16, and β values ranging from 0.4 to 0.5 (Figure 4.4). These results are in broad agreement with those obtained from later heather studies made at other sites (Table 4.2).

4.3 GRASS DRAINAGE LYSIMETERS—PLYNLIMON

Six monolith drainage lysimeters were installed in the grassland Wye catchment at Plynlimon—two on the tops of hills on peat soils at an altitude of 520 m; two on the slopes, on shallow podzolic soils overlying shattered bedrock; and two in the valley bottom, altitude 375 m, on peat soils overlying boulder clay. The fibreglass lysimeter tanks were of the type developed by the Agricultural Research Council's Letcombe Laboratory (since closed) and were installed using their equipment. The steel bases were initially sealed *in situ* using a resin compound; a more reliable seal was later achieved by welding, *in situ*, 15-cm deep 'sides' onto the baseplate and then 'potting' the tray so formed with a waterproof resin sealant. Drainage was aided by a ceramic candle set into the base of the tank, and a copper pipe carried the drainage water to a reservoir set in a pit beneath the tanks. To complete the installation a ground level raingauge was installed close to each of the tanks.

Neutron probe soil moisture observations were taken, using the method described by Bell (1976) and Calder (1979b), through centrally located access tubes so that the water balance of the

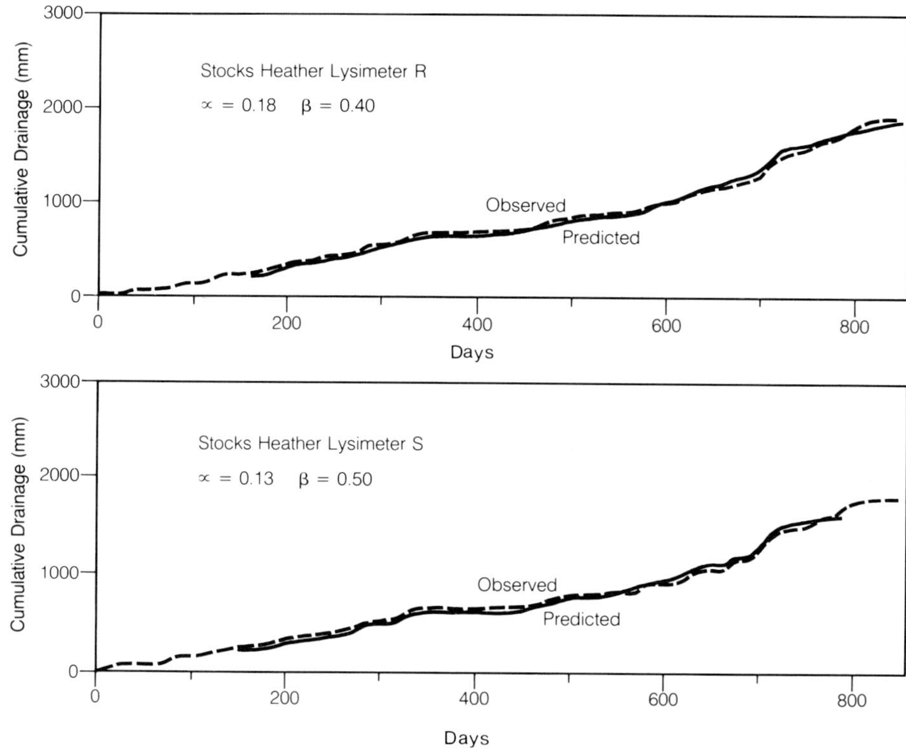

Figure 4.4 Observed and predicted cumulative drainage from the Stocks heather drainage lysimeters using data from 1965, 1967 and 1968 but excluding snow periods and periods when the data was of poor or uncertain quality.

tanks could be made over short time periods (usually 3–7 days). These measurements also provided a check on the integrity of the drainage results from the tanks.

During the winter, deficiencies in the operation of the lysimeters were evident; in certain periods, particularly those in which snowfall was recorded, more drainage than precipitation was recorded; this was particularly apparent in the more exposed hill-top lysimeters. Presumably undercatching by the raingauge or overcatching caused by the slightly raised tops (about 2 cm) of the tanks was responsible; whichever was true, the results could not be used to give a meaningful indication of winter losses.

Table 4.2 Summary of data from the Sneaton Moor heather lysimeter. (Reproduced by permission from Wallace *et al.*, 1982.)

Period (day no./year)	Length (days)	Rainfall (mm)	Evaporation (heather)(mm)	Penman E_T (grass) (mm) ($\approx 0.91E_T$ (heather))
35/80–35/81	365	992	473	494
35/81–35/82	365	981	478	504

During the summer months the operation of the lysimeters was more successful; the highests altitude tanks (tanks 1 and 2) and one of the lowest altitude tanks (tank 6) all gave continuous and good quality data from 14 May 1979 to 1 November 1979. The results from this selected period, of the cumulative depth of evaporation ΣE, were calculated from the cumulative water balance equation:

$$\Sigma E = \Sigma P - \Sigma D - \Delta S \qquad 4.1$$

where:

P = depth equivalent of precipitation
D = depth equivalent of drainage
S = depth equivalent of water content

and are shown in Figure 4.5. Also shown in Figure 4.5 are cumulative values of the daily Penman E_T estimate calculated from a nearby automatic weather station at Cefn Brwyn (altitude 359 m).

The measured evaporation during the summer months is, for all the tanks, lower than the Penman estimate. In these relatively high upland areas there is reason to expect that low temperatures may be restricting the growth of the grass during the spring and even during the summer months. Anderson and Harding (personal communication 1987) found that better model predictions were obtained with soil moisture data in Norway and the UK if a temperature-dependent regulating function, τ, was employed, which acted by setting predicted transpiration to zero when mean daily temperatures were less than 5 °C, setting it equal to E_T when temperatures were greater than 10 °C and reducing it proportionately for temperatures in between. Application of this function to the Penman E_T estimate obtained from the Cefn Brwyn automatic weather station data, using temperatures recorded at Cefn Brwyn and at a higher altitude (524 m) automatic weather station at Eisteddfa Gurig (also within the Wye catchment), resulted in significantly reduced evaporation estimates (Figure 4.5), which are broadly in agreement with the measured evaporation losses from the respective tanks at similar altitudes.

4.4 SOIL MOISTURE STUDIES OF TRANSPIRATION

Soil moisture observations were made under heather moorland and young and mature forest on the catchments of the Crinan Canal reservoirs, Strathclyde Region, Scotland (Calder *et al.*,

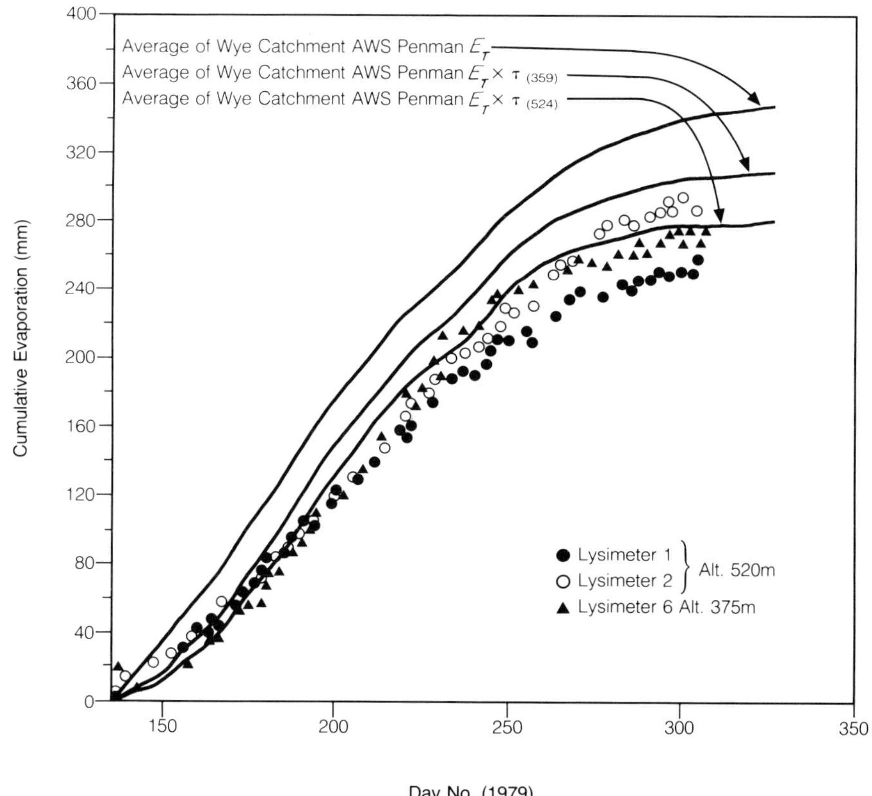

Figure 4.5 Measured cumulative evaporation from the Plynlimon grass drainage lysimeters together with evaporation estimates calculated from (1) Penman E_T (Wye catchment average value), (2) $E_T \times \tau$, where τ is calculated from temperature measurements at 359 m altitude and (3) $E_T \times \tau$, where τ is calculated from temperature measurements at 524 m altitude.

1982a; Calder, 1986), and under heather and grass moorland and mature forest on the Institute of Hydrology's experimental catchments at Balquhidder, Central Region, Scotland, and at Plynlimon, Central Wales.

Soil moisture modelling studies carried out using the Plynlimon observations (Calder *et al.*, 1983b) were consistent with evaporation rates given by the Penman E_T estimate and an infinite value of the root constant, which implied that even during the dry summer of 1976 the soil moisture deficits were not significantly affecting transpiration rates in this upland region.

At Crinan soil moisture deficits beneath heather are relatively small compared with those under forest (Figure 4.6), and modelling studies show that best agreement with observation is achieved when transpiration from the heather is assumed to take place at half the Penman E_T rate. The forest soil moisture deficits at both Crinan and Plynlimon are well represented by the daily

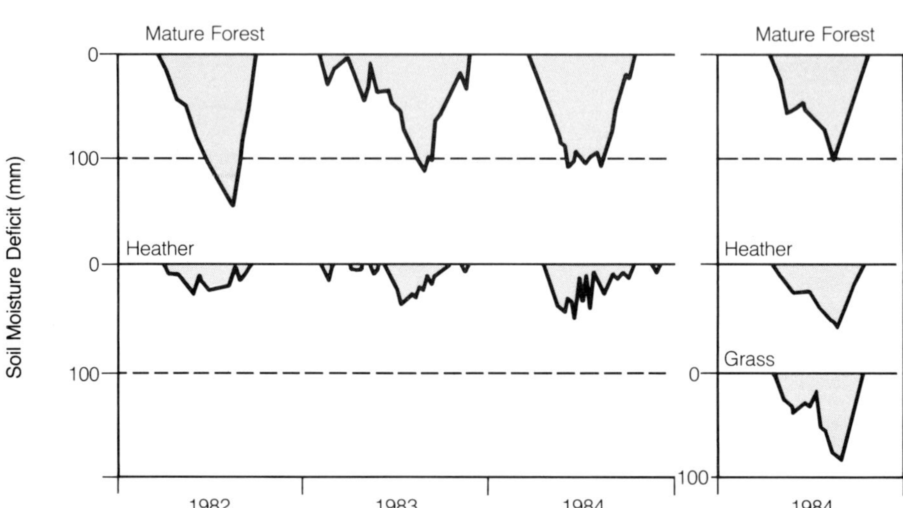

Figure 4.6 Soil moisture deficits measured beneath mature spruce forest, heather and grass moorland at Crinan and Balquhidder. (Reproduced by permission from Calder, 1986; Copyright Elsevier Science Publishers B.V.)

evaporation model (see section 9.4) when transpiration is assumed to occur at 0.9 of the E_T rate.

4.5 THE SNEATON MOOR HEATHER STUDIES

Perhaps the most detailed studies of transpiration from heather (*Calluna vulgaris* (L.) Hull) in the British uplands were those carried out by Roberts and colleagues of the Institute of Hydrology on the extensive heather tracts of the North Yorkshire Moors (Wallace *et al.*, 1982, 1984). The vegetation consisted of a pure stand of heather, which had not been burned for 15 years and, as a consequence, had formed a dense, complete canopy with a mean depth of 0.44 m. The site, at an altitude of 260 m, was on an east-facing slope of 5.3°; the surrounding topography was of similar open, gently rolling moorland.

Lysimeters were used, again of the Letcombe type but with load cells beneath so that they could be weighed. Other instrumentation comprised neutron probes, to give soil moisture information, and diffusion porometers and pressure bombs to measure stomatal resistance and leaf water potential. The changing climatic conditions were monitored with an automatic weather station, and eddy correlation instruments were used to investigate the wind and momentum transport regimes.

These studies showed that, during the measurement period, even in this drier east coast environment, soil moisture deficits generally were not limiting transpiration; the maximum soil moisture deficit recorded was 45 mm and the lowest soil water potential was -10 kPa. Transpiration rates were, nevertheless, very low compared with the Penman potential transpiration, E_T (heather) estimate. For dry days in May, June and July 1980 the ratio of actual to potential transpiration (heather) was 0.23, 0.32 and 0.46 respectively.

To express the transpiration as a ratio of the Penman E_T (grass) estimate, which has been used as an index of atmospheric demand in simple evaporation models (see Chapter 9), it was necessary to derive an approximate conversion factor. This was based on consideration of the different albedos of heather (≈ 0.13) and grass (≈ 0.25) whilst bearing in mind that the Penman E_T (heather) estimate made use of net radiation measurements made above heather. Thus, to a good approximation it is reasonable to expect that both the available energy and the E_T estimate for heather will be about 10% greater than those for grass, and the equivalent β values thus become 0.25, 0.35 and 0.50. The annual water balance obtained with the lysimeters is shown in Table 4.2.

The Sneaton Moor studies also provided results on the interception characteristics of heather. The aerodynamic resistance to vapour transport from the heather surface, measured by the lysimeter method (equation 5.3, see section 5.2), was found to be consistent with an inverse dependence with windspeed and the best fit was obtained with the relation:

$$r_a = 96u^{-1}(\pm 46\%) \qquad\qquad 4.2$$

Interestingly, for reasons which are still not understood (see section 11.3), the transfer resistance, calculated by applying a 'bluff body' correction to the aerodynamic resistance to the momentum estimate, derived from the eddy correlation instrument, gave a significantly larger value, $164u^{-1}$ ($\pm 17\%$), than that derived using the lysimeter.

CHAPTER 5

Rainfall Interception from Short Vegetation—The 'Wet Surface' Lysimeter Experiment

5.1 EXPERIMENTAL DESIGN

The studies outlined in the previous chapter provided basic information on transpiration from the major upland vegetation types; some of them also gave information on interception losses, but not in sufficient detail to allow the water retention characteristics or the aerodynamic transfer resistances to be determined. To obtain this information for short vegetation a portable 'wet surface' lysimeter system was designed in which the mass of a sample of wet vegetation, wetted either naturally by rain or artificially by spraying, was measured repeatedly. Following initial trials with a manually operated balance, an automatic system was constructed using a Sartorius 30 kg electronic balance connected to a Commodore PET microcomputer. At minute intervals, the microcomputer interrogated an automatic weather station (AWS), thermocouple and thermistor surface temperature sensors and CSIRO pattern soil heat flux sensors through an analogue to digital converter (3D, Warren St., London), and provided on-line calculation of both the evaporation rates and the aerodynamic transfer resistances (Figure 5.1). With uncommitted time between weather station scans the balance was interrogated at a rate approaching once per second and the readings averaged to reduce wind 'noise'. The sample tray (45 × 68 × 10 cm), which contained the vegetation sample, was constructed from glass fibre with an expanded polystyrene base which thermally insulated the sample from heat generated by the balance (Figure 5.2). To prevent flooding an electrically operated water pump, actuated by a float switch, was installed in the installation pit beneath the balance and tray assembly (Figures 5.3 and 5.4).

5.2 THEORY

The aerodynamic transfer resistances were calculated by two methods; firstly by inverse solution of the combination equation, for a wet surface (equation 2.6a):

$$r_a = \frac{\rho c_p (q_s(T) - q)}{\lambda E(\Delta' + c_p/\lambda) - \Delta' H} \qquad 5.1$$

and secondly by the definition, or Dalton equation (see section 2.2, equation 2.1a):

$$r_{aV} = \rho \frac{(q_o - q)}{E} \qquad 5.2$$

which, for a wet surface, is equivalent to

$$r_{aV} = \rho \frac{(q_s(T_o) - q)}{E} \qquad 5.3$$

Evaporation rates, E, were calculated from the changes of the averaged mass readings (using an 11-point moving average procedure) and the other variables in the equations were calculated from the meteorological and surface data. Because the different equations require different information, the uncertainties in the calculated aerodynamic resistances will be different and the magnitude of the difference will also vary with changing environmental conditions. A sensitivity analysis showed that, under virtually all meteorological conditions, accurate measurement of the surface temperature was crucial to the calculation of resistances using the definition equation (equation 5.2); under conditions when the atmosphere was close to saturation the analysis showed that large errors would result from the use of the inverse combination equation (equation 5.1).

5.3 SITES

The lysimeter system was operated at a number of locations in Scotland: at a site near Crinan on heather and grass; at a site near Killin on heather (Figures 5.5 and 5.6); and at Moormore near Aviemore on heather. The system has also been operated in England, at Berner's Heath near Thetford, East Anglia, on heather (Table 5.1).

From the onset of the experiments at the Crinan site the measurements clearly demonstrated the large (at the time surprisingly large) differences in evaporation rates which exist

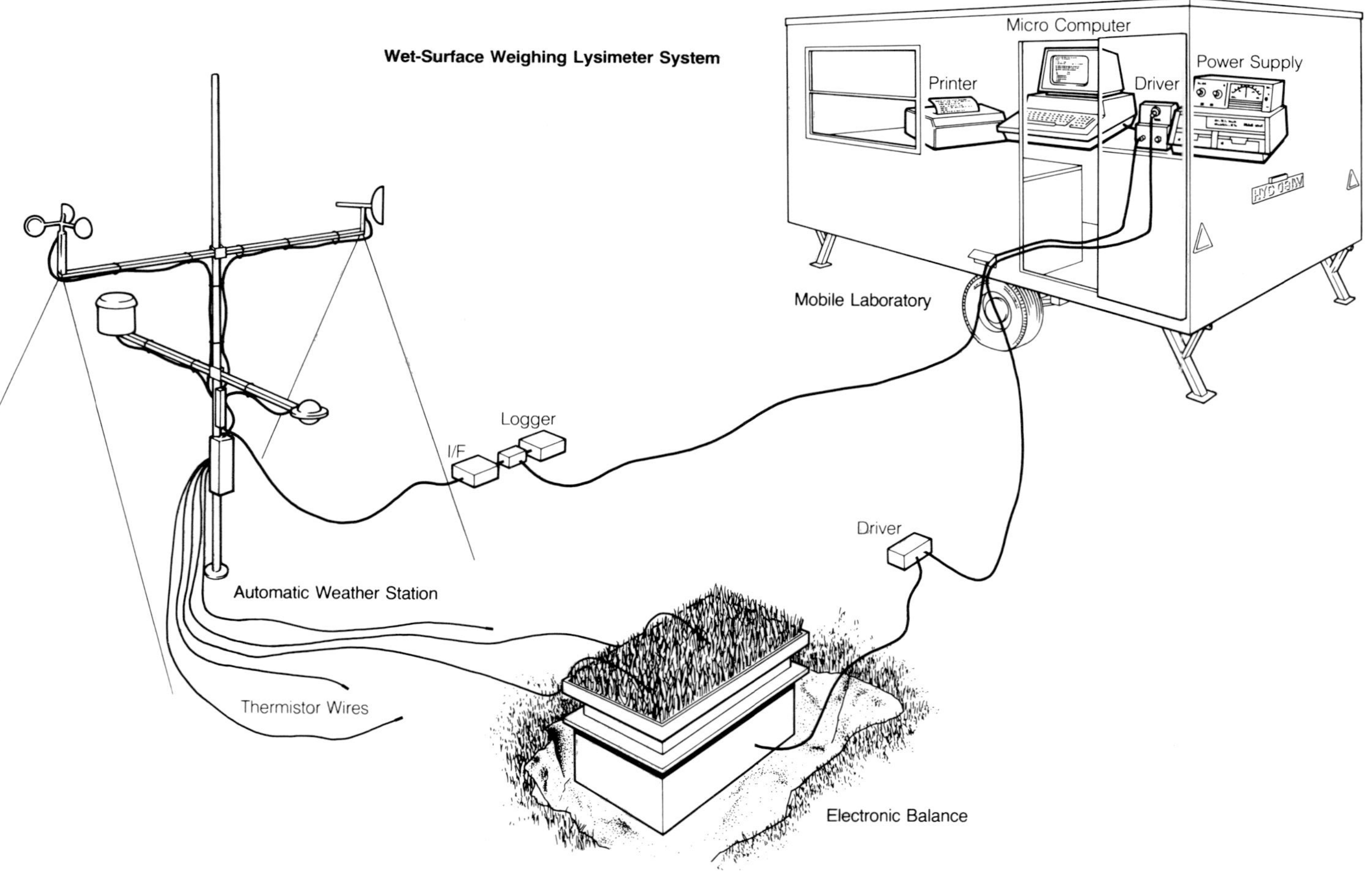

Figure 5.1 Schematic diagram of the wet surface weighing lysimeter system.

Figure 5.2 The heather sample in place in the lysimeter tray.

Figure 5.3 The electronic balance in its waterproof container being prepared for installation. The carrying handles, which are removed before installation, are also shown together with the breather pipe and the hose, which is used for dewatering the installation pit.

Table 5.1 Vegetation characteristics and values of the parameters C and i, which determine aerodynamic resistance (via the equation $r_a = Cu^i$, where u is the windspeed) at the different experimental sites

Site	Age	Height (m)	Ground cover (%)	LAI	P–M eqn.		Definition eqn.	
					C	i	C	i
Berner's Heath	6 yr	0.32±0.09	90±5	1.8±0.3	68±11	−1.3±0.3	74±8	−1.1±0.1
Moormore	Early building phase	0.33±0.13	85±5	1.8±0.5	44±9	−1.1±0.2	51±8	−1.0±0.1
Killin								
A	Mature	0.33±0.13	65±5	1.9±0.9	20±2	−0.6±0.06	33±2	−0.8±0.05
B	Mature	0.33±0.13	65±5	1.9±0.9				

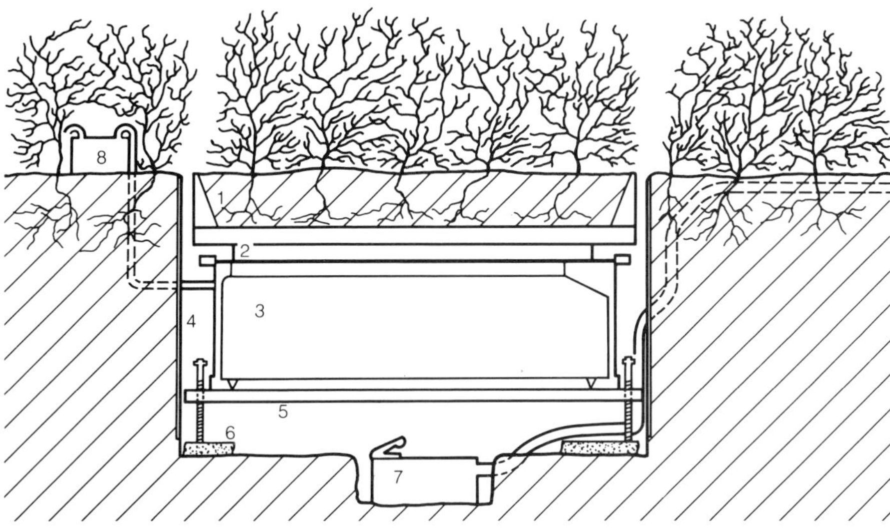

Figure 5.4 A cross-sectional diagram of the lysimeter showing: 1. the heather sample overlying an insulating block of polystyrene; 2. polystyrene spacer; 3. electronic balance in waterproof housing; 4. aluminium hole liner; 5. levelling tray; 6. concrete foundation blocks; 7. water pump; 8. breather box containing silica gel (Reproduced by permission from Calder *et al.*, 1984a; Copyright by American Meteorological Society.)

Figure 5.5 The wet lysimeter system in operation at the site near Killin, Scotland. The lysimeter is located to the right of the automatic weather station.

Figure 5.6 The data logging caravan in use with the wet lysimeter system at the site near Killin, Scotland.

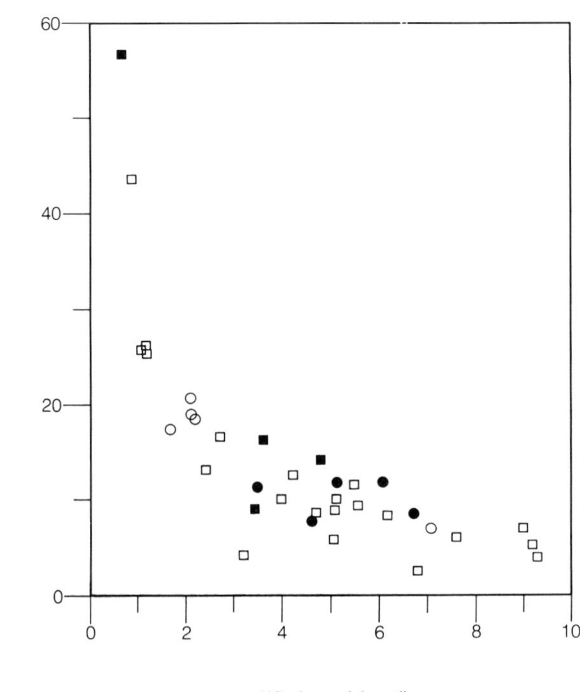

Figure 5.7 Aerodynamic resistance to vapour transport to a height of 1.2 m, r_{av}, as a function of the mean windspeed at a height of 2 m, calculated from equation 5.1 using data from the Killin heather sites. Key: □ natural rain event, site A; ○ spray event, site A; ■ natural rain event, site B; ● spray event, site B. (Reproduced by permission from Calder *et al.*, 1984a; Copyright by American Meteorological Society.)

between wet heather and wet grass when experiencing the same meteorological conditions.

5.4 AERODYNAMIC RESISTANCE RESULTS

Using the inverse Penman–Monteith equation (equation 5.1), the values of aerodynamic resistance, r_a, as a function of windspeed were calculated for the two sites (A and B) at Killin (Figure 5.7). The observations, made at the two sites in both rain and artificially wetted conditions, are differentiated by different symbols but statistical analysis has shown that there is no significant difference in the values in either of the two conditions (Calder *et al.*, 1984a). The power function which best fits all the observations is given by:

$$r_a = (32.6 \pm 2.0)u^{-(0.77 \pm 0.05)} \qquad 5.4$$

Generally the r_{aV} values calculated from the definition equation (equation 5.3) were greater than the r_a values calculated from inverse solution of the Penman–Monteith equation (equation 5.1). For less than half the observations the difference was less than one standard error. It is not known to what extent these small differences are due to real differences in the transfer resistances for heat and vapour or to experimental uncertainties; the r_a value calculated from the inverse Penman–Monteith equation is influenced by both transfer resistances, whilst the value given by the definition equation relates solely to the vapour component. The major experimental uncertainty concerns the estimate of the surface temperature, and this can arise either from the spatial variability of surface temperature or bias due to differences between the surface characteristics of the vegetation and the small (1 mm diameter) thermistor sensors. Until further evidence is available to the contrary it is therefore assumed that $r_a = r_{aV}$.

Differences between sites in the aerodynamic resistance–windspeed function are illustrated by differences in the sub-parameters C and i, which relate aerodynamic resistance to u, the windspeed at 2 m above the ground, via the equation:

$$r_a = Cu^i \qquad 5.5$$

The values of C and i were determined by linear regression of $\log(r_a)$ on $\log(u)$ using r_a values calculated from both the definition and inverse Penman–Monteith equations (Hall, 1987a) (Table 5.1). The aerodynamic resistance–windspeed functions obtained for the different sites using the inverse Penman–Monteith equation are shown in Figure 5.8. For windspeed less than 5 m s^{-1}

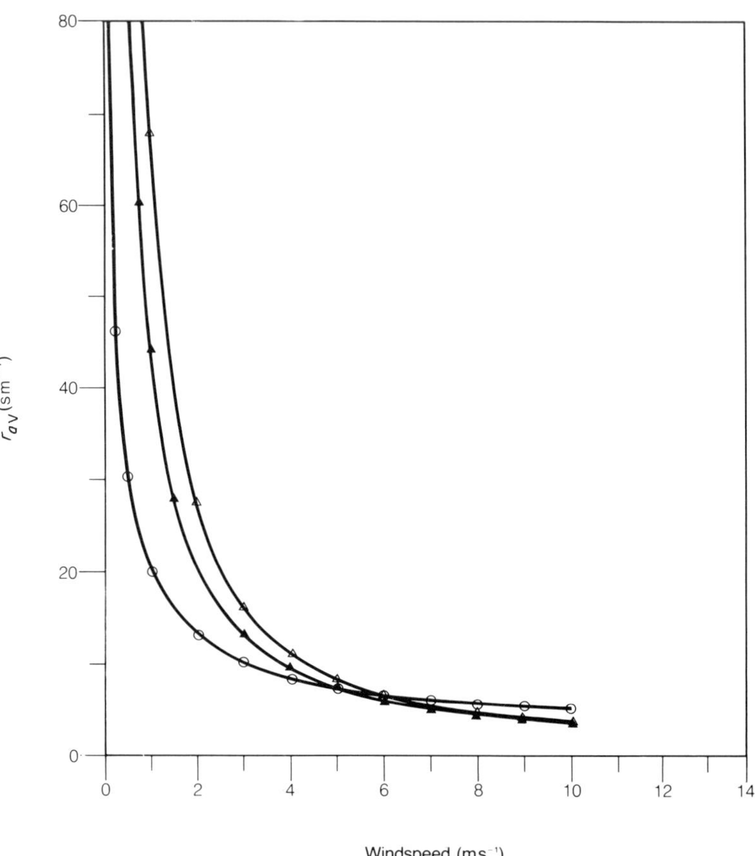

Figure 5.8 Aerodynamic resistance to vapour transport to a height of 1.2 m, r_{aV}, as a function of the mean windspeed at a height of 2 m, calculated from equation 5.1 using data from the different heather sites. Key: $\triangle$, denote the function $r_{aV} = 68u^{-1.3}\ sm^{-1}$, fitted to r_{aV} values for Berner's Heath; $\blacktriangle$, denote the function $r_{aV} = 44u^{-1.1}\ sm^{-1}$, fitted to r_{aV} values for Moormore; $\bullet$, denote the function $r_{aV} = 20u^{-0.6}\ sm^{-1}$, fitted to r_{aV} values for Killin. (Reproduced by permission after Hall, 1985; Copyright by Elsevier Science Publishers B.V.)

the resistances at Killin are less than those at Moormore, which in turn are less than those at Berner's Heath. The cause of the differences is not known but because the sites with lower resistances also have rougher topography it is possible to speculate that differences in atmospheric turbulence resulting from differences in the local topography may be an important factor. The Killin site was topographically the roughest, being located on a hillside in steeply mountainous terrain; the topography at Moormore was less severe, the site being located on a gently sloping valley bottom which was adjacent to mountains at distance of the order of a few kilometres; the Berner's Heath site was exceptionally flat, the surroundings being level and flat in all directions for distances exceeding tens of kilometres. Studies of the mechanism involved, whereby large-scale topographic features may be enhancing aerodynamic turbulence which, in turn, may be influencing the local aerodynamic transport mechanisms, are urgently required to understand this important feature of the interaction between evaporation and the atmosphere.

5.5 CANOPY DRAINAGE MEASUREMENTS

Although the 'wet-surface' lysimeter in the form described above is capable of giving very accurate measurements of evaporation for use in the calculation of aerodynamic transfer resistances, it is not capable of giving information on the retention characteristics of water on the canopy. Hall (1985) modified the method to allow this information to be obtained. The vegetation samples were prepared by first inserting an aluminium drainage gutter into the soil in the centre of the sample; a surrounding mould was then placed around the sample and molten paraffin wax was poured onto the soil surface until it was covered and sealed to the drainage gutter. The soil beneath the samples was then excavated and, by tilting the samples and adding more molten wax, a 'V'-shaped wax surface profile was obtained (with the gutter resting in the base of the 'V'). The sample and retaining wax base were trimmed of soil and placed within a drainer tray, which was specially constructed to prevent runoff from flowing beneath the sample (Figure 5.9). The whole assembly was then lowered onto the electronic balance within the pit in the usual manner and drainage experiments were begun immediately. These

Figure 5.9 The heather sample (AS3), contained within a waterproof wax block, in place on the drainer tray for use in canopy drainage studies.

usually continued for a period of a few days, and during this time no visually detectable canopy fall or deterioration of the samples occurred. (No detectable deterioration was evident even over time periods as long as two weeks.) The adopted procedure was similar to that for the evaporation measurements except that measurements were always carried out at night to minimise evaporative losses. Furthermore, particular attention was paid to measuring the mass of the sample both prior to and during spraying, as well as during the following drainage period.

By repeating this procedure, but by spraying beneath rather than above the canopy, the drainage characteristics of the wax base and drainer tray were determined independently of the sample. However, to resolve the very rapid changes in drainage rate from the plain wax surface, mass readings from the balance were recorded more frequently, at 1.2 s intervals.

The decrease with time of the amount of water held on sample (AS3) for two consecutive drainage experiments is shown in Figure 5.10. Even though there was a large difference in the recorded windspeed, 2.8 m s^{-1} during the first experiment and 4.8 m s^{-1} during the second, the results for the two experiments are, within experimental error, identical. There is therefore no

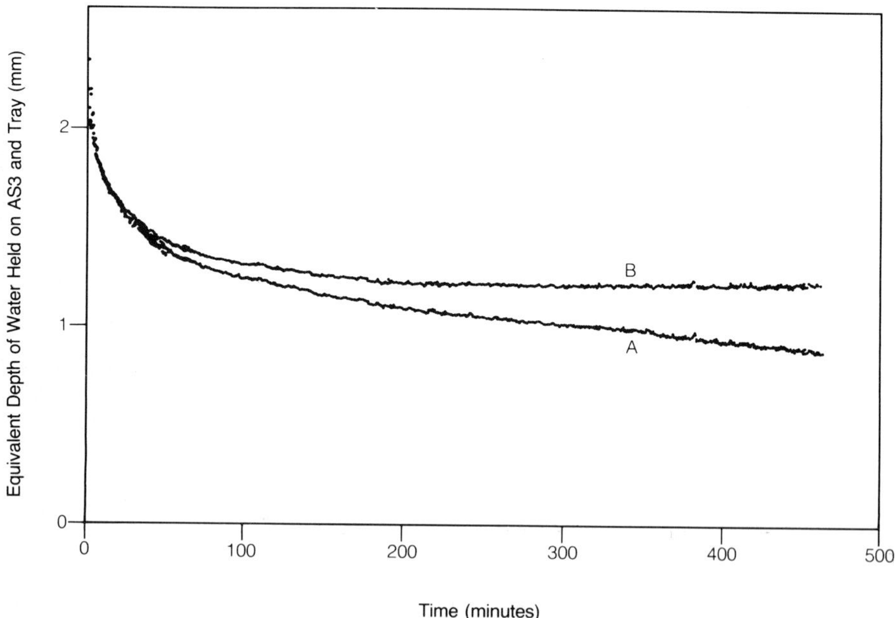

Figure 5.10 The change with time of the equivalent depth of water held on the heather sample (AS3) for two drainage experiments, curve A. The same data, after compensation for evaporative loss, showing the effects of drainage only, curve B. (Reproduced by permission after Hall, 1985; Copyright by Elsevier Science Publishers B.V.)

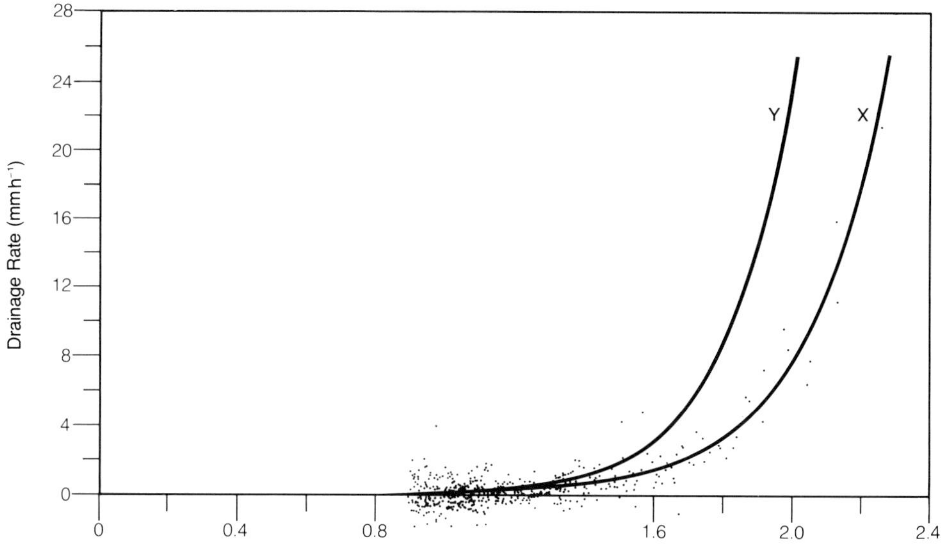

Water Held on the Heather Canopy, Wax and Tray, c_T (mm)

Water Held on the Heather Canopy, c (mm); Curve Y only.

Figure 5.11 Drainage functions calculated using the drainage rates shown in curve A of Figure 5.9. The drainage function for the heather sample and tray, $D_T = 0.00133$ $(\exp(4.31c_T - 1)$, which was fitted to the data points is shown as curve X. The drainage function derived for the heather sample alone, $D = 0.00085$ $(\exp(5.13c) - 1)$, is shown as curve Y. (Reproduced by permission after Hall, 1985; copyright by Elsevier Science Publishers B.V.)

evidence either for forests (see section 6.9) or heather of a windspeed dependence on the drainage function. On removal of the effects of evaporation, calculated from the Penman–Monteith equation using the Berner's Heath r_a value (Table 5.1), the change in weight of the sample due to drainage alone was determined and is shown in Figure 5.9 as curve B. All significant drainage ceased when the total storage reached a value of 1.2 ± 0.05 mm.

The drainage rates, calculated from curve B, as a function of canopy storage are shown in Figure 5.11 together with the fitted function

$$D_T = 0.00133(\exp(4.31c_T)-1) \qquad 5.6$$

where c_T = total depth per unit ground area of the water held on the vegetative sample, the wax and the tray.

On removal of the effects of the detention of water on the wax and tray using the relationship:

$$D_W = 0.021(\exp(27.09(c_W-0.1))-1) \qquad 5.7$$

where c_W = equivalent depth per unit ground area held on the wax and tray, the drainage relationship for water held on the heather canopy was obtained:

$$D = 0.00085(\exp(5.13c)-1) \qquad\qquad 5.8$$

This relationship is shown as curve Y in Figure 5.11.

5.6 EVAPORATION IN PARTIALLY WET CONDITIONS

The influence of partial wetness of the canopy on evaporation rates was determined by comparing measured evaporation rates E^M in these conditions with the Penman–Monteith estimate, E_W^{PM}, using an aerodynamic resistance given by equation 5.5. The results of a number of experiments in which values of the ratio of E^M and E_W^{PM} are compared (a) when the ratio was first observed to decline and (b) during the drying phase (Hall, 1984), is shown in Figure 5.12(a); the error bars represent a visual estimate of the average spread of the data. The mean value of canopy storage at which the decrease in the E^M/E_W^{PM} ratio occurred, was found to be 0.85 ± 0.1 mm (Figure 5.12(b)).

5.7 WETTING CHARACTERISTICS OF VEGETATION

The wetting characteristics of heather, spruce and grass leaves were investigated experimentally using a laboratory rainfall simulator. By using a very dilute tracer solution of rhodamine B the location of drops was clearly revealed, and recorded photographically, under ultraviolet illumination.

For heather it was found that only a relatively small amount of water was required to fully wet the canopy with a thin film, although significantly more was required before drainage was induced. The retentive nature of the heather surface is probably due to the lack of epicuticular wax, the intricate small-scale structure and the high wettability of the cuticle (Figure 5.13a) (Hall, 1985).

For spruce the water retention characteristics are intermediate between grass and heather. Only during continuous simulated rainfall was the water retained as a continuous film over the surfaces. On the cessation of rainfall it contracted into drops which lodged between the leaves and the leaf stalks (Figure 5.13b). The epicuticular wax that occurs on the upper (adaxial) surface of spruce leaves (Jeffree *et al.*, 1971) is probably responsible for this behaviour; little bridging of water droplets

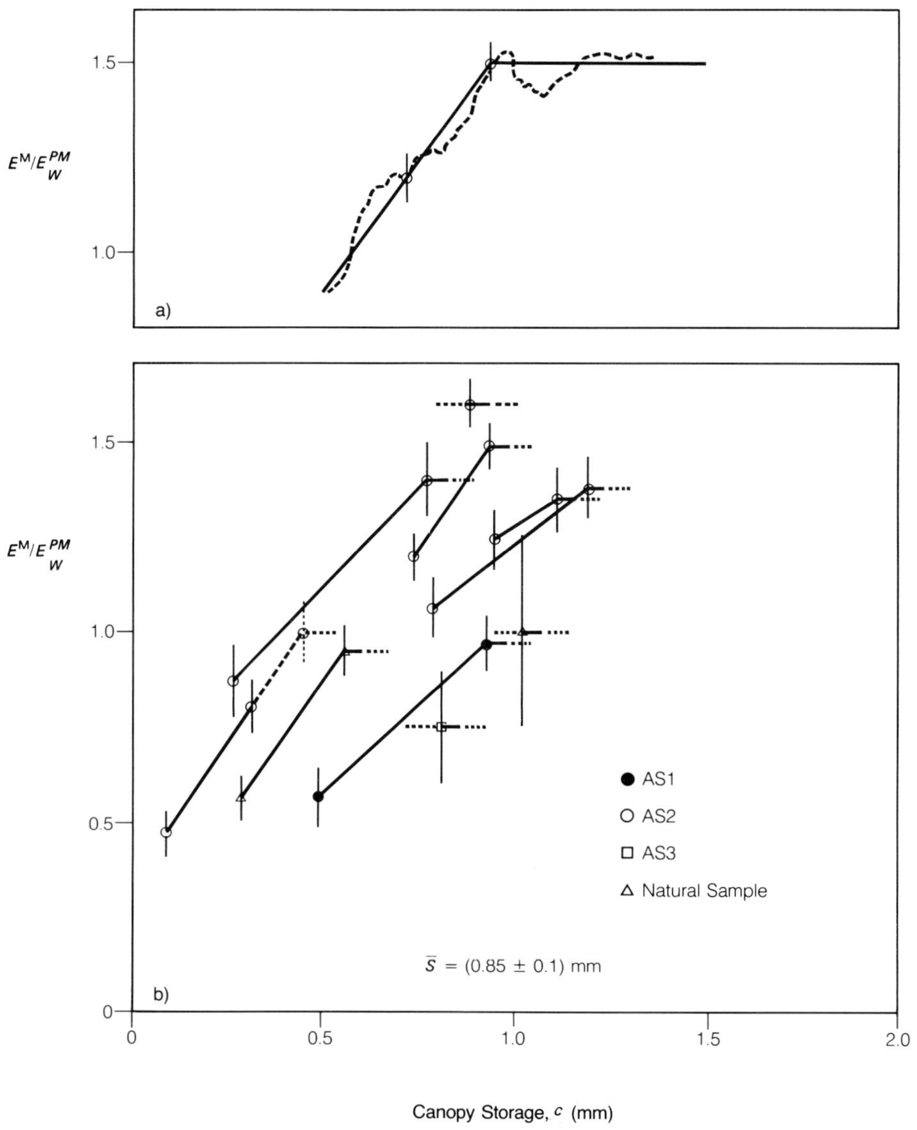

Figure 5.12 (a) Values of the ratio of E^M/E_W^{PM} plotted against canopy storage to illustrate the method of calculating the 'turning point' values and the 'average partially wet values' shown in Figure 5.12b.
(b) The 'turning point' and the 'average partially wet' values of the ratio of E^M/E_W^{PM} plotted against canopy storage obtained for a series of experiments using four different heather samples. (Reproduced by permission after Hall, 1985; Copyright by Elsevier Science Publishers B.V.)

between leaves occurs because, unlike heather, the spacing is too great.

When the same studies were repeated with grass samples the observed reduction in the wetting of the leaf surfaces was even more extreme. Water was retained only as small droplets, even

Figure 5.13 Photographs of water, containing an ultraviolet tracer, on samples of (a) heather, (b) spruce and (c) grass vegetation.

during precipitation, and only a small proportion of the leaf area was wetted at any time (Figure 5.13c).

The importance of differences in the surface wetting characteristics of different vegetation types has only recently been recognised. For spruce forest the evidence presently available suggests that evaporation rates reach the maximum E^M/E_W^{PM} ratio

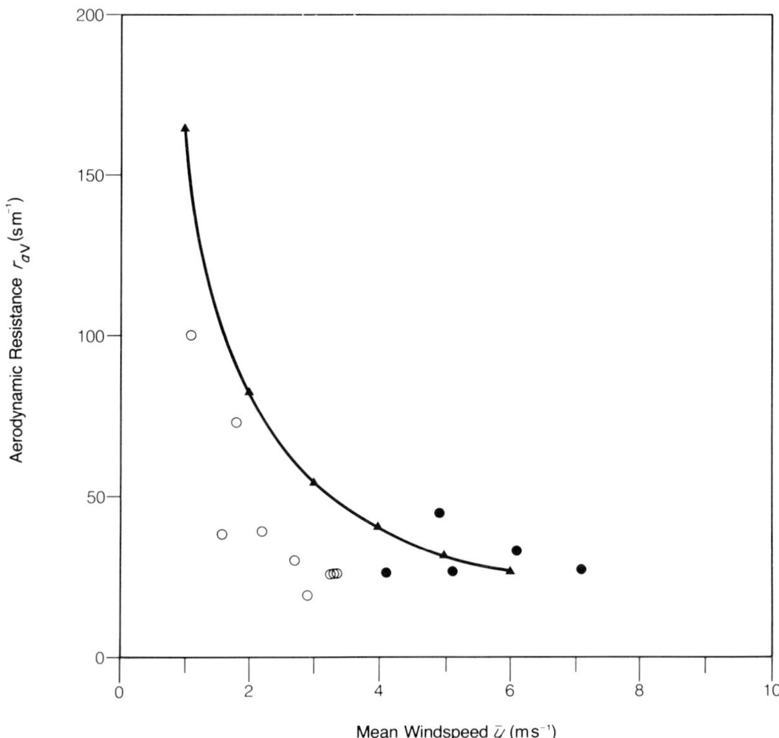

Figure 5.14 The aerodynamic resistance from the surface to a height of 1.2 m, r_a, as a function of the mean windspeed at a height of 2 m, calculated from equation 5.1 using data from experiments with: (a) surfactant-sprayed grass at Wallingford, denoted by open circles; (b) rain-wetted grass at Balquhidder, denoted by closed circles and calculated from classical eddy diffusion theory (equation 2.25) using the parameter values: k = 0.41, z = 2 m, d = 0.075 m and z_o = 0.01 m. (Reproduced by permission after Hall, 1985; copyright by Elsevier Science Publishers B.V.)

when the canopy capacity value of about 2 mm is reached (section 6.8) whereas for heather the evidence suggests that the E^M/E_W^{PM} ratio reaches a maximum when canopy storage exceeds 0.85 ± 0.1 mm (Figure 5.12) (Hall 1985), a value which is less than the canopy capacity value of 1.1 ± 0.1 mm at which significant drainage commences.

Under natural conditions it is unlikely that the leaves of living grass will ever be completely wetted, and evaporation rates will be less than the corresponding E_W^{PM} value. However, when a weak solution of surfactant is present to reduce surface tension forces, a continuous film can form and increase the wetted leaf area. Figure 5.14 shows the effective r_a values obtained when grass was wetted under natural conditions at Balquhidder and the generally lower values, for the same windspeed, found at Wallingford when the grass was sprayed with surfactant solution (Hall, 1987b).

CHAPTER 6

Gamma-Ray Attenuation Studies of Rainfall Interception from Forests

6.1 INTRODUCTION

Conventional methods for measuring interception loss from forests, which rely on comparisons between measured values of precipitation above and below the canopy, are not sufficient, when used on their own, to allow detailed investigation or verification of the processes involved. When used in conjunction with other methods for determining either the canopy storage or the evaporation rate much greater progress is possible.

Rutter (1966) recorded the change in weight of wet excised shoots to obtain independent measurements of the evaporation rate and, on a slightly larger scale, Hancock and Crowther (1979) used displacement transducers to record the weight of whole branches *in situ*: increasing the scale yet again Harding and Rosier (1984), adapting a method developed by Roberts (1977), used load cells to measure the change in weight of whole severed trees. To a greater or lesser extent, each of these methods is subject to the limitations imposed by sampling size and spatial representivity. In principle at least, this limitation can be reduced using a γ-ray attenuation method for determining canopy densities, from which canopy storage and evaporation rates can be inferred. The method is based on the measurement of the attenuation of a beam of γ-rays, by both photoelectric absorption and Compton scattering, in its passage through material situated between a radioactive source and a detector; a knowledge of the total attenuation coefficient for the material in the beam is also required for the successful application of this method.

6.2 EXPERIMENTAL DESIGN

The principal consideration in the design of a γ-ray attenuation system is the minimisation of the error dρ (or relative error dρ/

ρ) in the measurement of density subject to safety and cost constraints. The error arises largely from a combination of: (1) statistical random counting errors; (2) errors that are proportional to the count rate (for example, errors due to changes in detector efficiency); and (3) zero errors due to zero drift in amplifiers and analysers. With modern electronics zero drifts are probably insignificant compared with the other two sources of error.

These errors can be calculated from consideration of the attenuation equation, which states that if a homogeneous absorber of density ρ and attenuation coefficient μ is placed between a point source of γ-rays and a detector at distance r from the source, the flux of γ-rays observed at the detector is given by:

$$I_r = \frac{I_1 \exp(-\mu\rho r)}{r^2} \qquad 6.1$$

where I_1 is the unattenuated flux at unit distance from the source. The sensitivity of the detector, S_r, at distance r is, by definition, the negative of the ratio of the relative change in the attenuated flux and the relative change in the density of the absorber (for an infinitely small change in density) ie.:

$$S_r = -\frac{\mathrm{d}I_r/I_r}{\mathrm{d}\rho/\rho} \qquad 6.2$$
$$= -\frac{\rho\,\mathrm{d}I_r}{I_r\,\mathrm{d}\rho}$$

which, on differentiating equation 6.1 and rearranging gives:

$$S_r = \mu\rho r \qquad 6.3$$

From equations 6.2 and 6.3 the relative error in density determination is given by:

$$\frac{\mathrm{d}\rho}{\rho} = -\frac{(\mathrm{d}I_r/I_r)}{\mu\rho r} \qquad 6.4$$

The error in the count rate due to statistical counting errors is, if 'dead time' effects in the analyser and rate scaler can be neglected, equal to the square root of the count rate and thus:

$$(\mathrm{d}\rho/\rho)_{r,\text{Statistical}} = \frac{(I_1\exp(-\mu\rho r)/r^2)^{1/2}\,/(I_1\exp(-\mu\rho r)/r^2)}{\mu\rho r}$$
$$= -\frac{\exp(\mu\rho r/2)}{\mu\rho r(I_1)^{1/2}} \qquad 6.5$$

For errors which are proportional to the count rate, i.e. $dI_r = cI_r$, the relative error in the density determination is given by:

$$(d\rho/\rho)_{r,\text{Proportional}} = \frac{-c}{\mu\rho r} \qquad 6.6$$

Random counting errors can be reduced by increasing source intensity or counting times. However, long counting times will increase the effects of 'drifts' (changes in efficiency in the detector system with time) and increase the value of c. The best compromise is to arrange that the random counting error and proportional errors are of a similar magnitude; this was adopted in the design of the system.

6.3 EXPERIMENTAL METHOD

6.3.1 The γ-ray system

The γ-ray attenuation system used in the upland evaporation studies was originally developed by the Applied Physics Department of Strathclyde University (Olszyczka, 1979). This system was configured so that a beam of 660 keV γ-rays, emitted by a 200 millicurie caesium-137 radioactive source, traversed a horizontal distance of tens of metres through the forest canopy before striking a detector, comprising a 0.3 m² plastic scintillator attached to a photomultiplier. The source and detector, suspended from towers, were winched up and down manually to allow the beam to scan different levels in the canopy.

Subsequent developments were aimed at improving the stability of the detector and increasing the data acquisition rate by automating the system and bringing the positioning of the source and detector under computer control (Figure 6.1). The control computer also monitored, and could be configured to control, the temperature of the photomultiplier housing on the detector. It also measured the count rate of the pulses from the scintillation counter, calculated an appropriate dead time correction, and evaluated the mass per unit area (M) of material in the beam from solution of the attenuation equation:

$$M = \frac{1}{\mu}\log((n - n_b)/(n_a - n_b)) \qquad (\text{kg m}^{-2}) \qquad 6.7$$

where:

μ = attenuation coefficient for water (0.008564 m² kg⁻¹; Hubbell, 1969)

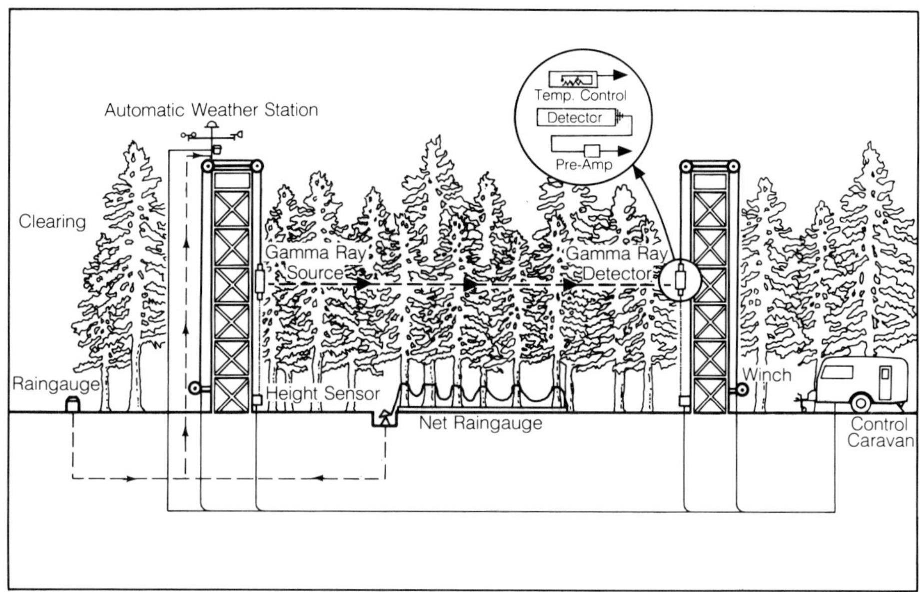

Figure 6.1 Schematic diagram of the forest γ-ray attenuation system (not to scale).

n = attenuated count rate
n_a = count rate with only air in the beam
n_b = background count rate.

6.3.2 Calibration and experimental errors in the density determination

Proportional errors in the count rate, when the system was standardised by reference to count rates in air, were found to be less than 1%. The relative random counting error with the chosen 20 second counting time was 0.3%, giving a total relative error in the density determination of 1.1%. This represented an absolute density error at an individual scan level typically of 0.03 kg m^{-3}. If the errors at each level are assumed to be uncorrelated, the error in the measurement of total mass per unit ground area will then be approximately 0.1 kg m^{-2} giving an error in the measurement of intercepted water of 0.1 mm depth.

A calibration experiment was also conducted to measure the resolution and accuracy of the system and to investigate its general integrity. This involved placing plywood sheets normal to the standard 'air count' beam positioned at a height of 2 m above the ground surface and midway between the source and detector. Each sheet had dimensions of 2.44 × 1.22 × 0.01 m and a mean weight per unit area of 5 kg m^{-2}. The experiment simulated in all respects the normal operation of the system except that the source and detector were held fixed. The reference count rate, with only air in the beam, was obtained initially, then, without moving either the source or the detector, the number of sheets in the beam was increased and the new attenuated count rates recorded. The sheets were then removed, the air count rate re-measured, and the mass per unit area of the sheets calculated from equation 6.7 with the operational computer analysis program.

The results of this procedure, which was carried out on two separate occasions, are shown in Figure 6.2. The calibration was found to be stable, reproducible and linear; a regression analysis indicated a value of 0.00859 m² kg⁻¹ for the attenuation coefficients of the plywood sheets. The attenuation coefficient of cellulose, the principal constituent of wood, is known to be 3% less than that of water (see Hubbell, 1969) so the experimentally determined coefficient is perhaps 1 or 2% higher than expected but is within the error introduced by the uncertainty in the weight, the precise chemical composition, and the spatial variation in density of the sheets.

To investigate the possibility of the calibration being sensitive to the position of the sheets in the beam, which might be expected if the primary beam contained a significant number of secondaries scattered from the vegetation or the ground, the sheets were also inserted at intermediate positions, both closer to and further away

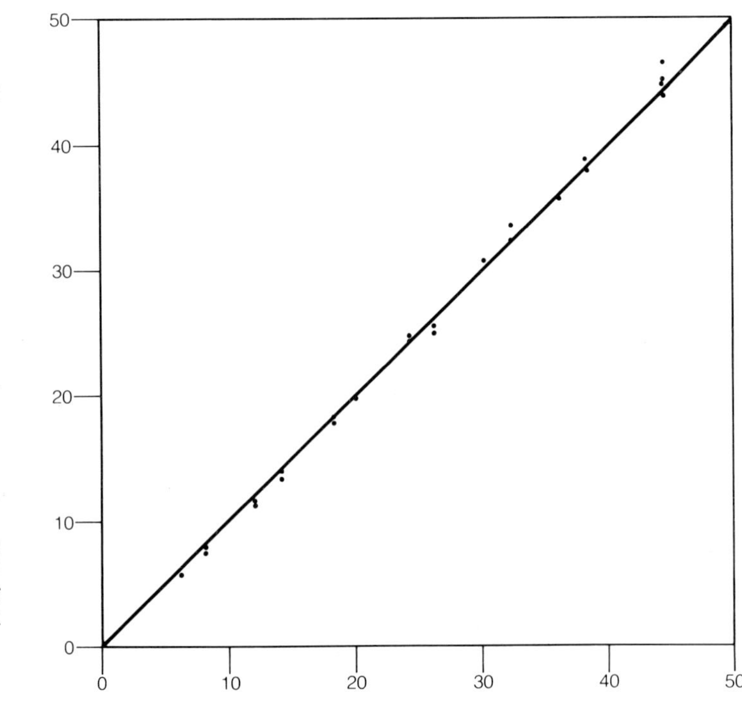

Figure 6.2 Gamma-ray determination (μ = 0.008564 m² kg⁻¹) of the mass per unit area normal to the beam of plywood sheets plotted against gravimetric values. (Reproduced by permission from Calder and Wright, 1986; Copyright by the American Geophysical Union.)

from the source. No significant differences were found. In all subsequent analyses the attenuation coefficient for water (0.008564 m^2 kg^{-1}) was used; as a result a small systematic error of 1–2% may have been introduced into the calculation of dry canopy density, for which no correction was made.

6.3.3 On-line calculation

The computer also interrogated an automatic weather station sited on the top of the source tower, and obtained information on rainfall and net-rainfall from a 0.1 mm resolution tipping bucket raingauge sited in a forest ride close to the site, and a 0.025 mm resolution plastic-sheet net-rainfall gauge sited beneath the canopy (Figure 6.1). From this data the computer performed on-line calculations of canopy density, drainage rates and the Penman–Monteith estimate of evaporation in wet surface conditions when the surface resistance (r_s) may be equated to zero, E_W^{PM} (equation 2.23a).

6.4 DESCRIPTION OF SITES

The improved automated system was originally constructed and tested at the Dolydd office site in the Hafren Forest in mid-Wales and observations were made in wet conditions. The system was then moved to the Queens Forest site, near Aviemore in Scotland. At both sites the trees were mature sitka spruce, of a similar age, tree density and height. Details of the site characteristics are given in Table 6.1.

At the Hafren Forest site the beam was directed midway between and parallel to the original lines of tree planting. This arrangement had the advantage that beam alignment could be

Table 6.1 Forest site characteristics

	Hafren Forest Site	Queens Forest Site
Location	Longitude 3°40′W	Longitude 3°42′W
	Latitude 52°28′N	Latitude 57°11′N
Altitude (m)	310	345
Tree species	Sitka Spruce	Sitka Spruce
	(*Picea sitchensis* (Bong.) Carrière)	
Planting date	1950	1953
Maximum tree height (m)	14.5	15
Mean diameter at breast height (cm)	45	46
Tree density; stems per hectare	3900	3100
Beam length from source to detector (m)	40	40
Beam length through forest (m)	25	35

set up and checked easily but had the disadvantage that the portion of the forest canopy scanned by the beam, which lay midway between the tops of the trees, may have been unrepresentative of the canopy as a whole. A further disadvantage, which became apparent when the system was used in windy conditions, was that not only was random 'noise' introduced into the canopy density measurements as tree branches swayed within the beam, but the systematic increase was observed as the tree tops were displaced into the beam. At the Queens Forest site the beam was arranged to intersect both the canopy and the trunk of the trees, which ensured a more representative canopy sample. As was expected, errors in the measurment of canopy density were much reduced with this arrangement.

6.5 CANOPY DENSITY PROFILES

Density profiles of dry forest, measured at 1 metre height intervals, are shown for both sites in Figure 6.3. At the Hafren site the beam did not intersect any tree trunks so the density profile relates only to the branches and leaves in the canopy, and

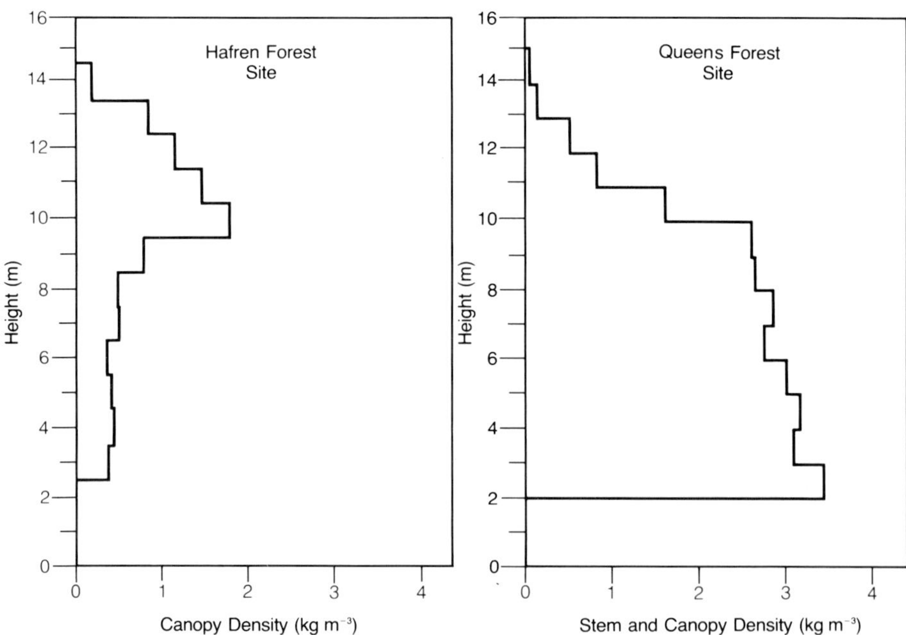

Figure 6.3 Density profiles under dry conditions of the forest canopy at the Hafren site and of the stems and canopy at the Queens Forest site. (Reproduced by permission from Calder and Wright, 1986; Copyright by the American Geophysical Union.)

a peak in the profile is observed at the level where the closure of the live canopy occurs. At the Queens Forest site the beam intersected both the canopy and three tree trunks, and the presence of the trunks accounts for the gradual increase in measured biomass density with decreasing height; the total density of canopy and trunks per unit area was 26.6 kg m^{-2}. (The measured trunk density was not representative of the forest as a whole as the proportion of tree trunks to horizontal beam area is less than that for the rest of the forest; at breast height the ratio of trunk cross-sectional area to ground area is 0.0019 in the beam as compared with 0.0052 for the surrounding forest.)

At the Hafren site the total mass per unit ground area of the canopy was found to be 8.0 kg m^{-2}. The measurements were obtained over the winter period from December to April, a period during which needle drop would have greatly exceeded new growth and the annual cycle of canopy density would have been expected to be close to the minimum. The measured canopy density value is therefore probably not inconsistent with the higher mean value of 11.0 kg m^{-2} obtained by staff of the Macaulay Institute, Aberdeen (reported by Olszyczka, 1979), who used harvesting techniques in six different closed canopy stands of sitka spruce.

6.6 WET CANOPY OBSERVATIONS

Changes in the total mass per unit ground area of the canopy, recorded during a typical storm event at the Hafren site, are shown in Figure 6.4. At 09.00 GMT the canopy was already saturated from overnight rain. Consistent increases in the total mass of the wet canopy were recorded following the onset of rainfall. These total mass values, on subtraction of the dry canopy mass per unit ground area (8 kg m^{-2}), represent the storage of water on the canopy, which can be expressed more conveniently as a depth of water (mm). Figure 6.4 also shows that when the canopy storage is in excess of a value of about 2.0 mm, appreciable net rainfall occurs; this canopy storage value represents what is often (rather crudely) termed the canopy capacity. A more exact description of the canopy drainage characteristics can be given in terms of a canopy drainage function.

Solution of the water balance equation allows evaporation rates from the canopy to be calculated from the difference between the measured rainfall and net-rainfall rates minus the rate of change of canopy storage. The cumulative value of the measured evaporation, E^M, together with that predicted by the Penman–Monteith equation, E_W^{PM} (r_{aM} = 3.5 s m^{-1}, determined from the Plynlimon forest lysimeter studies; Calder, 1977) is also

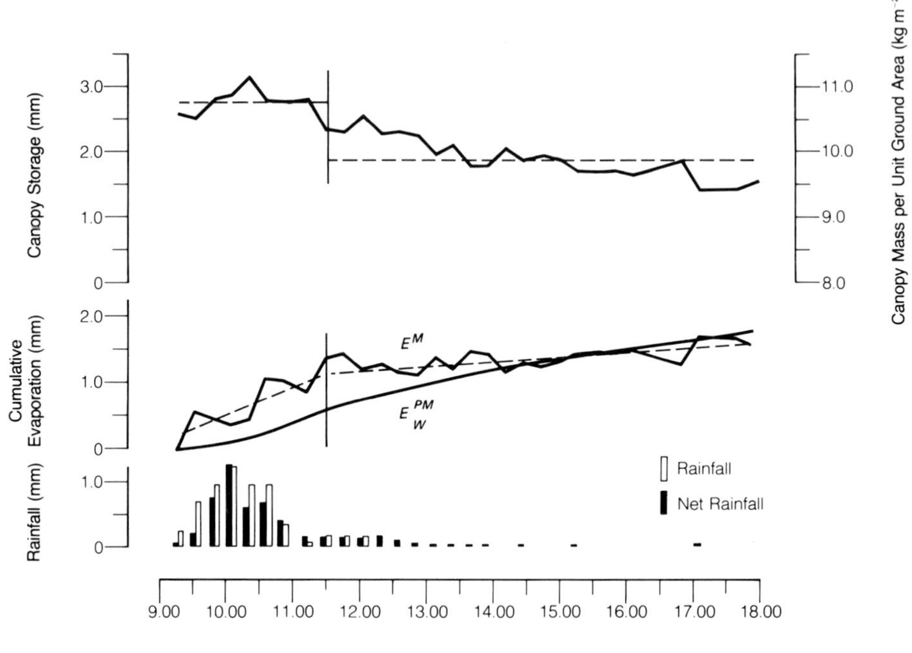

Time (hours GMT) 17 March 1983

Figure 6.4 Rainfall, net-rainfall, canopy storage, measured evaporation rate, E^M, and calculated evaporation rate, E^{PM}_W (r_a = 3.5 s m^{-1}) plotted against time. Dashed curves indicate the average values of c and E^M over the time periods selected for the subsequent calculation of evaporation rates. (Reproduced by permission from Calder and Wright, 1986; copyright by the American Geophysical Union.)

shown in Figure 6.4. Both the measured and predicted values demonstrate similar trends, with higher evaporation rates during the early part of the event, reducing to lower rates later. The absolute magnitudes of the predicted and measured evaporation rates were 0.28 and 0.38 mm h^{-1} for the first 2 hours respectively, when the canopy was fully wetted, and 0.18 and 0.07 mm h^{-1} respectively for the remaining 6 hours, when the canopy was partially wet.

6.7 CANOPY DRAINAGE FUNCTION

Six different events were recorded in which appreciable drainage occurred, and the relationship between drainage rate and canopy storage for these events is shown in Figure 6.5. The trends demonstrated by each of the events recorded at both the Hafren and Queens Forest sites are similar, with drainage rates rapidly

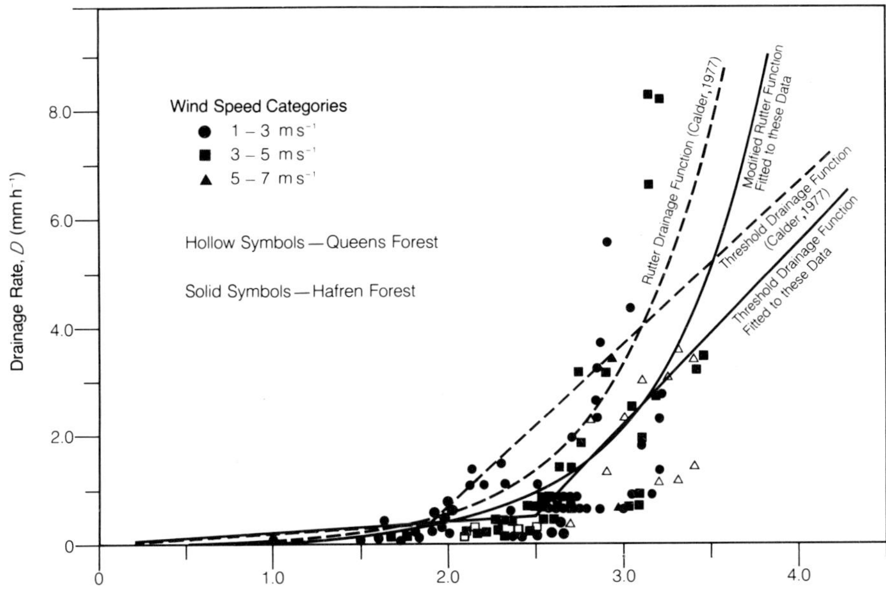

Figure 6.5 Net-rainfall intensity measured with plastic-sheet net-rainfall gauges at the Hafren and Queens Forest sites plotted as a function of canopy storage. Modified Rutter and threshold drainage functions, fitted to these data, are shown by the solid curves. Also shown for comparison are the Rutter and threshold drainage functions (dashed curves) which were derived by optimisation methods, using only rainfall, net-rainfall and meteorological data, from a different site in the Hafren Forest. (Reproduced by permission from Calder and Wright, 1986; copyright by the American Geophysical Union.)

increasing with increasing canopy storage when the canopy capacity value (2.0 mm) is exceeded. Also shown for comparison in Figure 6.5 are the two optimised drainage functions which were derived, again from the Plynlimon Forest lysimeter data, by fitting the predicted net-rainfall intensity from an interception model to that measured by the plastic-sheet net-rainfall gauges. These functions are given by:

$$D = 0.0018e^{1.76c} \qquad\qquad 6.8$$

using the Rutter model and

$$D = 0.21c \qquad\qquad c < 1.9 \text{ mm}$$
$$\text{or } D = 3.2c - 5.6 \qquad\qquad c \geq 1.9 \text{ mm} \qquad\qquad 6.9$$

using the threshold model described by Calder (1977).

6.8 EVAPORATION FROM PARTIALLY WET CANOPIES

The description of the evaporation process for conditions when vegetation is only partially wetted has not yet been adequately realised. Fortunately, for practical purposes an exact description is not essential for wet forests, as the calculated interception losses for a storm event are insensitive to the predicted partially wet evaporation rate, except where rain occurs before the canopy has completely dried. The description is nevertheless of great interest from a theoretical viewpoint and has engaged the attention of many workers; see, for example, Shuttleworth (1976, 1977), Monteith (1977), Massman (1980), Sellers and Lockwood (1981), and Hancock *et al.* (1983).

The results from the γ-ray system provided new insights into the mechanisms operating in partially wet conditions. Figure 6.6 shows the ratio of the measured to predicted evaporation rates, E^M/E_W^{PM}, in relation to canopy storage. The measured evaporation rates were calculated over time periods sufficient to reduce the

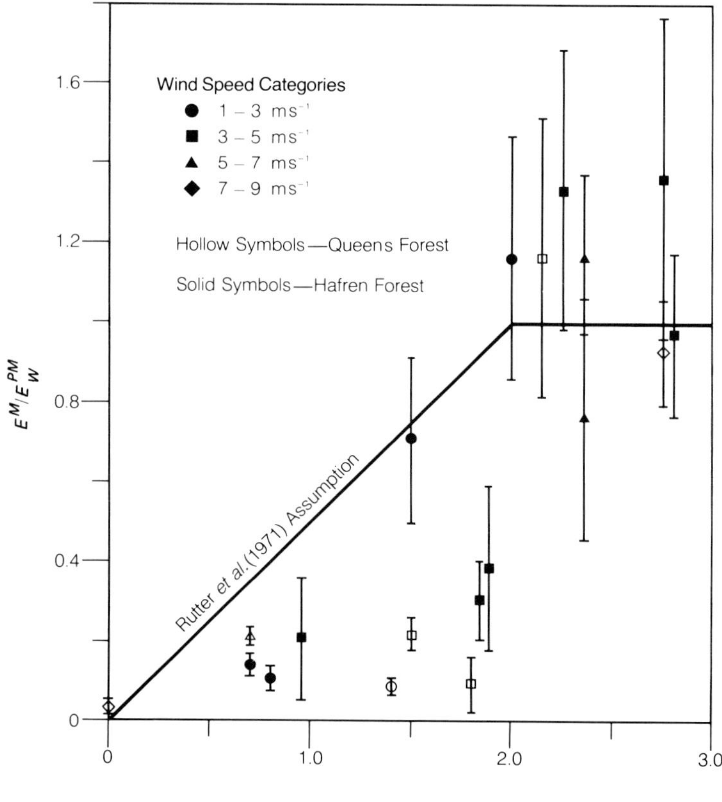

Figure 6.6 Measured evaporation rate, E^M, divided by calculated evaporation, E_W^{PM} ($r_a = 3.5$ s m^{-1}), plotted against canopy storage. (Reproduced by permission from Calder and Wright, 1986; copyright by the American Geophysical Union.)

proportional error to an acceptable value; these periods ranged from 1.5 to 7.5 hours. The ratio of E^M/E_W^{PM} reduces with decreasing canopy storage for values of storage which are less than 2 mm. Above this value the E^M/E_W^{PM} ratios are consistent with unity, an indication that the aerodynamic resistance value of 3.5 s m^{-1} employed in the calculation of E_W^{PM} is appropriate (the mean of these E^M/E_W^{PM} values weighted according to the inverse square of their error bars, is 1.07 ± 0.16). Figure 6.6 also shows that for storages less than the canopy capacity value the ratio of E^M/E_W^{PM} is less than that predicted by the linear relationship between evaporation rate and canopy storage that was assumed by Rutter *et al.* (1971). Application of the Rutter assumption would therefore tend to overestimate evaporation rates from partially wet canopies and underestimate the drying time for the lower levels of the canopy.

6.9 DEPENDENCE OF DRAINAGE PARAMETERS AND AERODYNAMIC RESISTANCE ON WIND SPEED

Wind speeds recorded during operation of the γ-ray system ranged from 1.5 to 8.0 m s^{-1}. Different symbols identify the wind-speed ranges associated with the canopy drainage rates versus canopy storage points shown in Figure 6.5. There is no indication from these results that drainage rates are influenced by wind speed. Clearly the modelling of forest interception losses is much simplified if wind-speed dependence can be discounted.

From consideration of conventional eddy diffusion theory (see section 2.8), the aerodynamic resistance to the transport of momentum in neutral conditions is given by equation 2.37:

$$r_{aM} = \frac{(\ln(z-d)/z_o)^2}{k^2 u(z)}$$

Rutter *et al.* (1975) assumed that for wet forests the aerodynamic resistance to heat and vapour transport could be calculated from this equation, with the roughness length and zero plane displacement calculated from mean tree height, h:

$$d = 0.75h \qquad\qquad 6.10$$

$$z_o = 0.10h \qquad\qquad 6.11$$

For the Hafren site, where the majority of the results were obtained, h was 13 m, and z, the height at which mean wind speed was measured, was 18 m. For the mean wind speed, $u =$

3.75 m s^{-1}, recorded during saturated conditions for the periods shown in Figure 6.6, the calculated aerodynamic resistance (from equation 2.37) is 5.4 s m^{-1}. Clearly, application of this formula would result in much lower values of E_W^{PM} and higher E^M/E_W^{PM} ratios which are generally more divergent from unity than those shown in Figure 6.6 (where r_a = 3.5 s m^{-1}). For the range of wind speeds observed, which vary by a factor of four, this theory would also predict that the values of E^M and E^M/E_W^{PM} would vary by a similar factor; even though the uncertainties in the E^M/E_W^{PM} ratios are large, they clearly could not encompass such a variation. Indeed, there is no indication that the scatter in the ratios is correlated with wind speed. A weighted linear regression of E^M/E_W^{PM} ratios against wind speed for saturated canopy conditions gave a slope of -0.035 s m^{-1} with an uncertainty of $\pm$ 0.036 s m^{-1} and a correlation coefficient of 0.36.

CHAPTER 7

Plastic-Sheet Net-Rainfall Gauge Measurements of Forest Interception

7.1 INTRODUCTION

The spatial variability of precipitation reaching the ground beneath forest canopies (termed net-precipitation or net-rainfall) is large when averaged over areas less than that occupied by a single tree. Measurements made with conventional 0.4 m^2 throughfall troughs and single tree stemflow gauges in the Hafren Forest at Plynlimon indicated coefficients of variation of 27% and 41% respectively. A large number of gauges are required to achieve a reasonable degree of accuracy with this method and this entails a high labour requirement, not only for the construction of the gauges but also, and more importantly, for the routine maintenance and cleaning of twigs and leaves from these gauges. Furthermore, if continuous flow measurements are needed, commensurately large numbers of flow gauges will also be required together with their associated logging equipment.

The large (40 m^2) plastic-sheet net-rainfall gauge (Figure 7.1) was developed for use with the Plynlimon Forest lysimeter (Calder and Rosier, 1976) to circumvent these difficulties and has subsequently been used successfully at many sites in the UK and also overseas, in tropical rainforests in Java (Calder *et al.*, 1987) (Figure 7.2) and Brazil (Shuttleworth *et al.*, 1984).

The basic design, in which heavy-gauge plastic sheeting is overlaid on a rope framework (Figure 7.3), to channel both drips from the canopy and flow down the stems of the trees into a plastic gutter and then into a large tipping bucket, has remained unchanged from that originally described by Calder and Rosier (1976). Small modifications involving the use of an isocyanate-based sealing compound rather than silicone rubber to seal the sheets to the stems of the trees has resulted in more durable seals; minor modifications have also been made to the waterproofing of

Figure 7.1 (a), (b) Plastic-sheet net-rainfall gauge used with the Plynlimon Forest lysimeter.

Figure 7.2 Plastic-sheet net-rainfall gauge used in studies of tropical rainforest evaporation, West Java.

the mechanical counters used with the buckets, and additional reed switches have sometimes been incorporated to allow an electronic digital counter to be attached in addition to a multichannel AWS logger.

The gauges used in the snow interception studies (section 8.2) were identical except for the addition of soil heating cables placed beneath them and around the tipping bucket flowmeters; the heating system was actuated by an 'ice alert' system, which is commercially available and was originally developed for de-icing road surfaces.

A dynamic calibration (Calder and Kidd, 1978) was used to calculate flow rates from the number of tips recorded in each 5-minute scan interval of the AWS logger. The dynamic calibration was necessary to take into account the small proportion of the flow that is 'lost' whilst the bucket is tipping and the central division of the bucket approaches the inlet stream. If the time taken to reach this position is t and the flow rate is such that the time between successive tips of the bucket is T, the dynamic flow calibration is:

$$Q = V/(T - t) \qquad\qquad 7.1$$

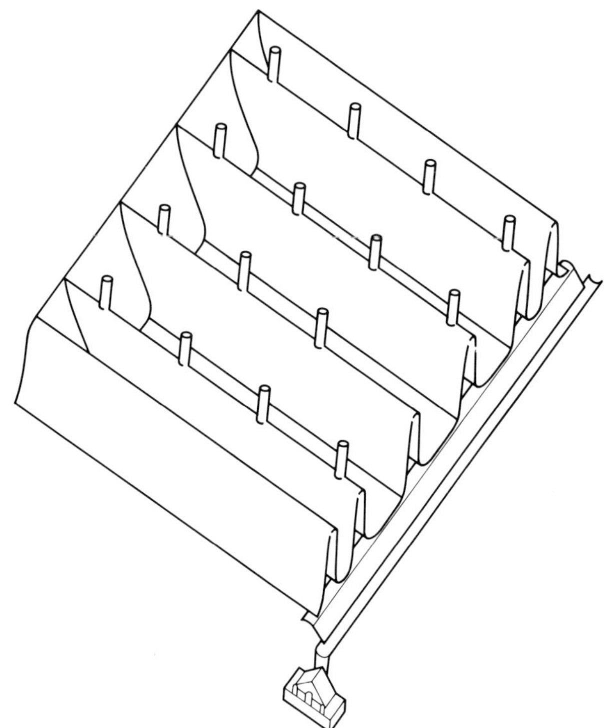

Figure 7.3 Outline diagram of a
plastic-sheet net-rainfall gauge.

where Q represents the flow rate and V is defined as the static
volume of the bucket. For both the large ($1.4\,\mathrm{dm}^3$) tipping
buckets used on these gauges and the small RIMCO type
raingauge tipping buckets the tipping time t is found to be about
$0.5\,\mathrm{s}$. On average, for sites in the UK where $40\,\mathrm{m}^2$ plastic-sheet
gauges have been used, the calculated flows are about 1% higher
than those calculated without the dynamic correction.

7.2 STUDIES IN WALES AND SCOTLAND

7.2.1 Measurements

Plastic-sheet net-rainfall gauges have been installed, always in
replicate pairs, in association with: (1) the Plynlimon forest
lysimeter; (2) the γ-ray attenuation experiment at the Dolydd
office site, Plynlimon; (3) the Crinan Canal reservoir resources
study in a forest on the Kintyre peninsula; (4) the γ-ray attenuation site in the Queens Forest near Aviemore.

The integration loss expressed as a percentage of the precipitation (α) recorded at these different sites, together with results
from other UK studies, is shown, plotted against the annual
rainfall rate, in Figure 7.4 (for observation periods which were

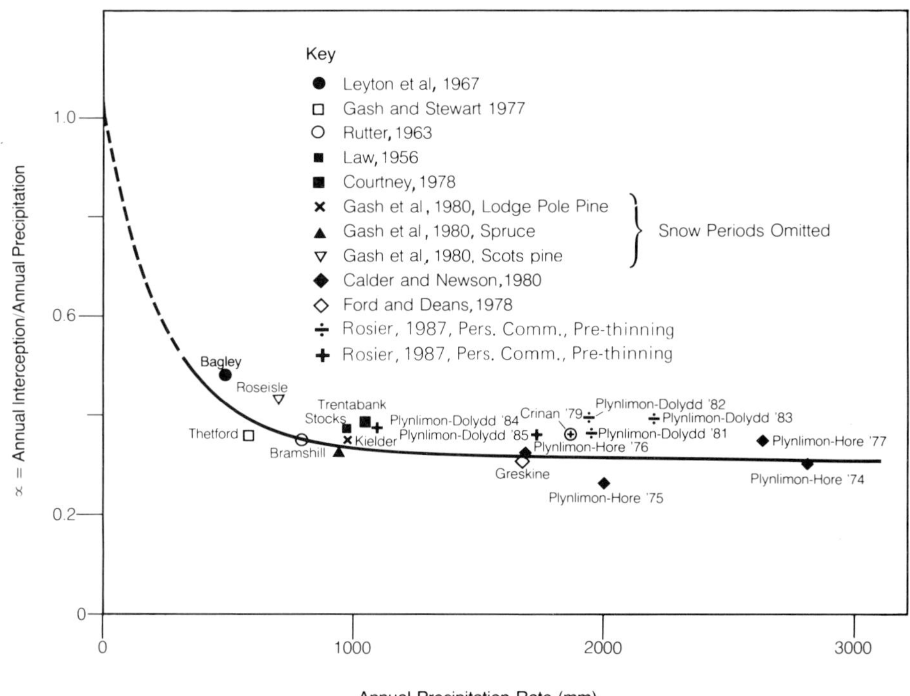

Figure 7.4 Observations of the annual fractional interception loss, α, from forests in the United Kingdom.

less than 1 year the annual rainfall rate was calculated for the observation period). For annual rainfall regimes exceeding 1000 mm the interception ratios are about 0.35. For all sites, except the Aviemore site, snow does not normally represent a large proportion of the annual precipitation. Observations made during snow periods have been included in the analyses except when observations indicated the possibility of errors due to the freezing of the buckets; in these circumstances the observations were discounted.

7.2.2 Experimental errors

The error in the measurement of net-rainfall arises primarily from uncertainties in the tipping-bucket flow calibration, which are known to be about 3%, and from errors due to the spatial variability of net-rainfall. From consideration of the variability measured at the Plynlimon-Hore site with 0.4 m^2 throughfall troughs (coefficient of variation = 27%) and stemflow gauges (coefficient of variation = 41%), the spatial variability error in the estimate of the true net-rainfall with a 40 m^2 gauge encompassing the area occupied by 100 troughs and including (say) 13 stems is (assuming stemflow accounts for 30% of the net-rainfall) about 3%. The total error is therefore expected to be less than 5%. The measurements from the replicate pairs, which generally diverge by about or less than this figure, give verification to this error calculation. Taking the mean of the measurements from two gauges should therefore result in errors in the estimate of the true net-rainfall of about 3.5%.

The assessment of errors in the measurement of rainfall is difficult. When rainfall measurements are made in forested, topographically steep, upland areas the task is even more formidable. If, as a working hypothesis, the error involved in using a ground-level gauge in a forest clearing is 5%, the error in the measurement of interception loss will be about 15%, (where interception loss is about 35% of the rainfall).

7.2.3 Site comparison

The cumulative rainfall values and interception losses measured at the different sites are shown in Figures 7.5, 7.6, 7.7 and 7.8. The effects of site differences on the model parameters, for the two-parameter exponential model,

$$I = \gamma(1 - \exp(-\delta P))$$

(where I is the daily interception loss, P is the daily rainfall and γ and δ are model parameters, see section 9.4), were investigated

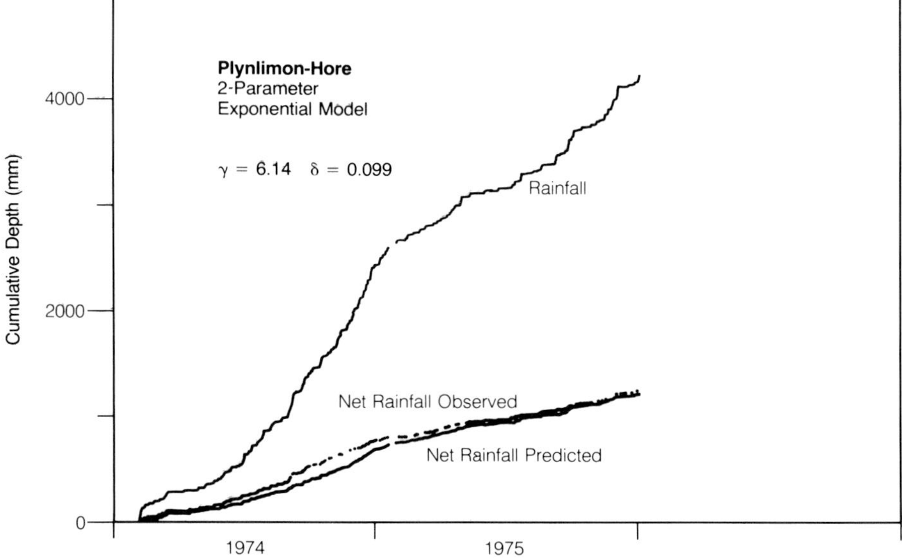

Figure 7.5 Measured cumulative rainfall and interception loss together with cumulative interception loss, fitted with the two-parameter daily exponential model, at the Plynlimon-Hore site.

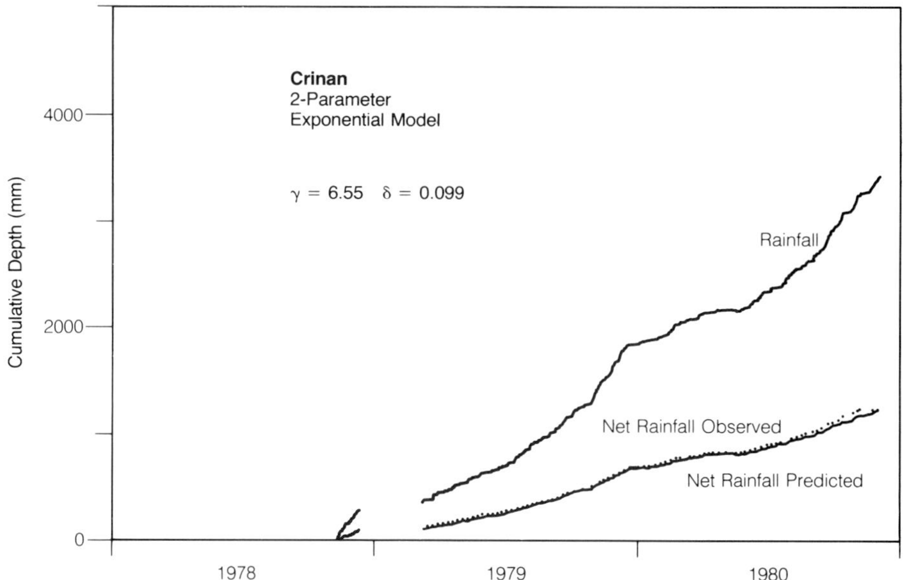

Figure 7.6 Measured cumulative rainfall and interception loss together with cumulative interception loss, fitted with the two-parameter daily exponential model, at the Crinan site.

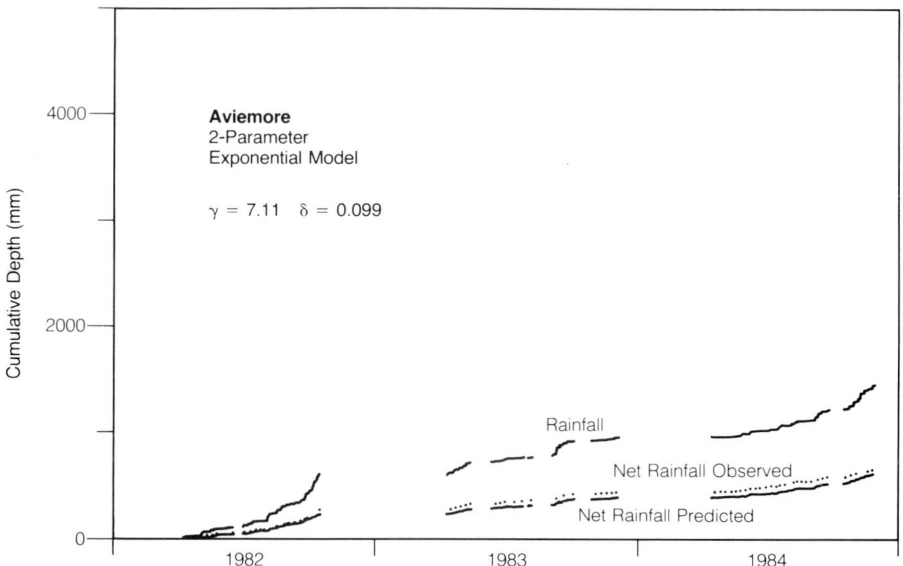

Figure 7.7 Measured cumulative rainfall and interception loss together with cumulative interception loss, fitted with the two-parameter daily exponential model, at the Aviemore site.

by keeping the parameter δ fixed at the value optimised over all sites (see section 9.4) and optimising the parameter γ. The optimised values of γ and the indices of goodness of fit are shown in Table 7.1; the model predictions of cumulative interception

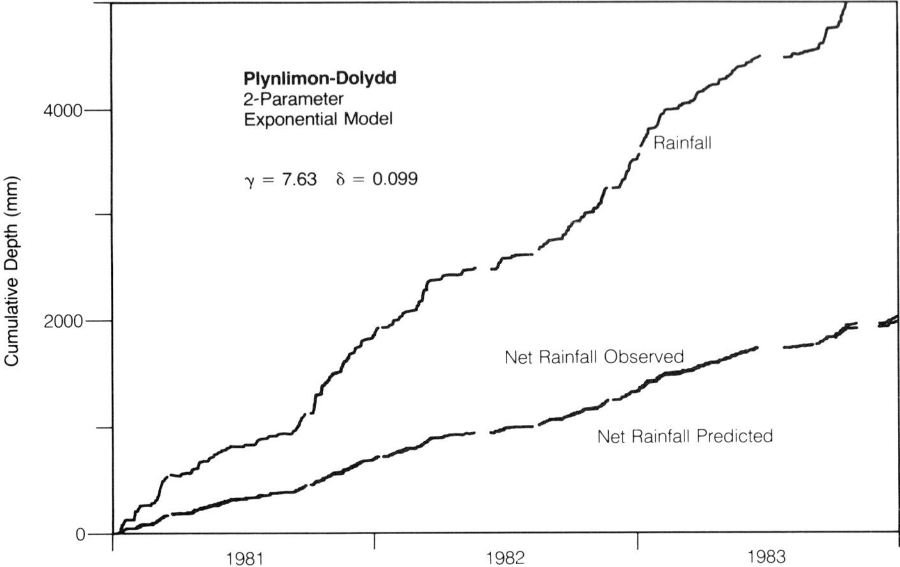

Figure 7.8 Measured cumulative rainfall and interception loss together with cumulative interception loss, fitted with the two-parameter daily exponential model, at the Plynlimon-Dolydd site.

loss are shown in Figures 7.5–7.8. Although the differences exhibited between sites are of a similar order to those expected from experimental error (15%) it is tempting, nevertheless, to speculate that the high values for the γ parameter exhibited at the Plynlimon-Dolydd site may be associated with the close proximity of the site to the edge of the forest. In the prevailing upwind direction, the site is within 250 m of the edge of the forest, whereas all the other sites are in excess of 1 km from the edge.

Table 7.1 Values of the optimised parameter, γ, ($\delta = 0.099$) and indices of goodness of fit obtained for the different interception sites and, for the Plynlimon-Dolydd site, both prior to and following thinning. The 'RMS error' refers to the relative root mean square error in the prediction of interception loss for 1-day, 30-day and 365 day periods

Site	Observation period	γ (mm)	Sum of squares of deviations (mm^2)	RMS error over period (days)		
				1 (%)	30 (%)	365 (%)
Plynlimon-Hore	1974–1975	6.15	1093	70.2	27.6	6.7
Crinan	1978–1980	6.54	404	50.4	25.5	3.0
Aviemore	1982–1984	7.12	101	40.8	18.5	11.4
Plynlimon-Dolydd (pre-thinning)	1981–1983	7.63	1308	49.2	14.4	2.9
Plynlimon-Dolydd (post-thinning)	1984–1985	6.94	403	52.3	15.3	0.6

7.3 TREE-THINNING EXPERIMENT—PLYNLIMON

Many interception experiments have been carried out in the UK but in the main these have been on complete and mature stands; few, if any, studies have been reported from either thinned or immature forest.

The opportunity for investigating the thinning aspect was provided when, after three years of observation on mature forest at the Plynlimon-Dolydd site, the normal forestry practice of line thinning, by removal of one line in three, was carried out.

The plastic sheets were reinstalled on their original locations after the thinning and observations continued for a further three years.

The interception ratio values obtained (Table 7.2) show little alteration following the removal operation; certainly they have not been reduced in proportion to the percentage of canopy removed.

To identify quantitatively the effects of the thinning, in isolation of possible effects due to the reduced rainfall in the post-thinning period compared with the period before, a modelling study was carried out.

Initially the daily 'two-parameter exponential' interception model was used. This was calibrated over the period prior to the thinning and predictions of interception loss were obtained over the whole measurement period, using the parameter values optimised for the pre-thinning period ($\gamma = 7.63$, $\delta = 0.099$), and compared with observations (Figure 7.9). Although after thinning the measured interception losses are less than those predicted, the differences are clearly small. The optimised value for the γ parameter decreased by 10% (Table 7.1), which indicates that

Table 7.2 Annual breakdown of rainfall and interception measurements made prior to and following thinning at the Plynlimon-Dolydd site

Year	Number of observation days	Measured rainfall (mm)	Measured interception (mm)	Interception ratio	Annual rainfall rate over observation period (mm a^{-1})
Pre-thinning					
1981	350	1862	676	0.363	1942
1982	324	1665	642	0.385	1876
1983	294	1768	686	0.388	2195
Post-thinning					
1984	263	790	294	0.372	1098
1985	365	1767	627	0.355	1767

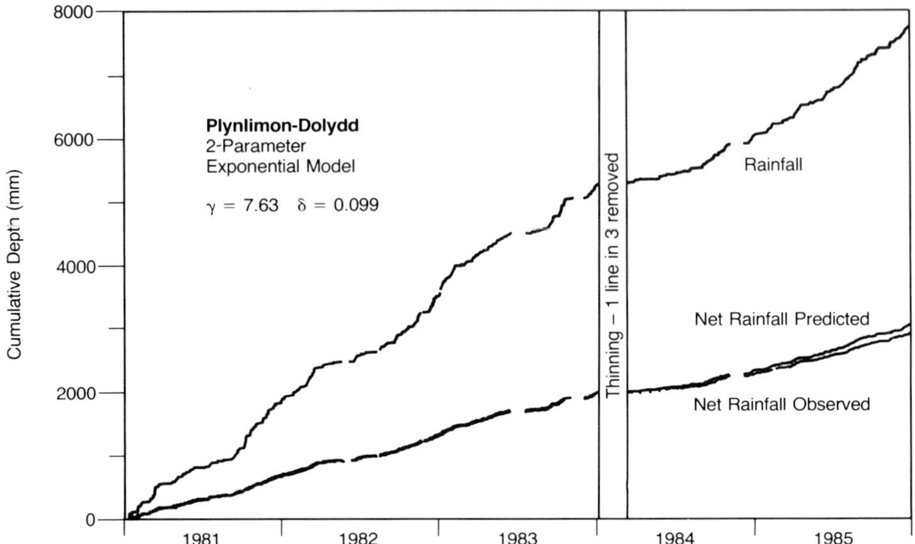

Figure 7.9 Measured cumulative rainfall and interception loss together with predicted interception loss, obtained with the two-parameter daily exponential model using parameters fitted during the pre-thinning period, at the Plynlimon-Dolydd site.

total interception losses, for any given rainfall pattern, would also decrease by this amount. Clearly this reduction is significantly smaller than the reduction of 33% which might be expected simply from pro-rata considerations of the reduction of canopy density.

A possible explanation for this result is that although canopy storages and leaf area index must decrease in direct proportion to the thinning, the increased ventilation to lower levels of the canopy may, notwithstanding the decrease in leaf area index, result in a lower aerodynamic resistance to the transport of water vapour and, consequently, rates of evaporation may actually increase. If this explanation is correct the duration of wet conditions following rainfall will be reduced as a result of both the reduction in canopy capacity and the enhanced evaporation rates.

Results from the application of the three-parameter stochastic model, both before and after thinning, lend some support to this explanation. Before thinning the parameter values were : $c_{max} = 4.05$ mm, $\bar{E}/\bar{R} = 0.13$ and $q = 1.08$ (Table 7.3). After thinning one-third of the forest stand it would be expected that the c_{max} value would be reduced in direct proportion to 2.64. From consideration of the increase from 0% to 15% in the free throughfall fraction, estimated from anascope measurements of the canopy coverage, q would be reduced by approximately 15% to 0.92 (see section 3.3). The $\bar{E}/\bar{R}$ ratio was then optimised, with

Table 7.3 Values of the optimised parameters. $\bar{E}/\bar{R}$, q and c_{max} and indices of goodness of fit obtained with the three-parameter daily stochastic model for the Plynlimon-Dolydd site, both prior to and following thinning. For the post-thinning period only the $\bar{E}/\bar{R}$ parameter was optimised; the other parameters were calculated from estimates of the effects of the thinning.

Site	Observation period	$\bar{E}/\bar{R}$	q	c_{max}	$\bar{R}$	$\bar{E}$	Sum of squares of deviations
				(mm)	(mm d^{-1})	(mm d^{-1})	(mm^2)
Plynlimon-Dolydd (pre-thinning)	1981–1983	0.126	1.08	4.05	8.08	1.02	1197
Plynlimon-Dolydd (post-thinning)	1984–1985	0.177	0.92	2.64	6.97	1.23	353

the expected parameter values fixed, for the post-thinning period. Good model fits (compared with the application of the exponential model, Table 7.1) were obtained and the optimised $\bar{E}/\bar{R}$ was greater than the pre-thinning value. This increase is partially due to the decrease in $\bar{R}$ in the post-thinning period but the $\bar{E}$ value has, nevertheless, increased by about 20% subsequent to thinning (Table 7.3).

Confirmation and further investigation of this somewhat surprising result is required. One approach would be to make simultaneous measurements of interception losses from both thinned and unthinned stands in the same locality, using plastic-sheet net-rainfall gauges, ideally in conjunction with γ-ray total canopy density measurements; results from this type of experiment should allow the effects of thinning to be clearly distinguished from the effects of any climatic variation.

CHAPTER 8

Evaporation from Snow-Covered Vegetation

8.1 EVAPORATION FROM A 'SMOOTH' SNOW PACK

When snow covers short vegetation the resulting surface generally becomes, aerodynamically, very much smoother. It would therefore be expected that the turbulent transport of both heat and vapour from the new surface would be inhibited. Furthermore, because the albedo of the snow surface is generally much higher and there is an extra energy requirement in terms of the latent heat of melting, it is reasonable to assume that evaporation rates will be much reduced as compared with a wet vegetated surface experiencing the same external meteorological conditions.

Experiments carried out at Finse, Norway, to study the exchange processes, mainly for improving snow melt models, have indeed confirmed this general view (Harding, 1986). The studies, which involved neutron probe measurements of the equivalent water content of the pack, the use of the portable lysimeter (see Chapter 5) to measure evaporation and condensation rates, and measurements of the temperature and wind profiles above the pack showed that evaporation rates never exceeded 0.06 mm h^{-1} and, over time periods of the order of 1 day, evaporative losses were generally counterbalanced by almost equivalent amounts of condensation.

Measurements from the lysimeter of aerodynamic resistance of the pack to water vapour transport and wind profile measurements of the aerodynamic resistance to momentum transport were usually consistent with the aerodynamic resistance function:

$$r_a = 300u^{-1} \quad (s\ m^{-1})$$

where u is the wind speed (m s^{-1}) recorded at a height of 2 m. (Resistances significantly higher than those given by this equation were, however, recorded from the very smooth surface produced immediately after snowfall; Harding, 1986). Despite the relatively high aerodynamic resistance of the snow surface, about half of the energy for melting the pack in spring was, nevertheless, supplied by sensible heat transport (usually during stormy periods), the rest being accounted for by radiant energy inputs.

8.2 SNOW INTERCEPTION FROM FORESTS— INTRODUCTION

The importance of snow interception has been the subject of considerable debate; the observation that snow accumulations beneath canopies are less (even when snow held on the canopy is taken into account) than in adjacent clearings has been interpreted in two ways: firstly, through snow redistribution in windy conditions from the canopy into the clearings and, secondly, by increased evaporation rates from snow intercepted by the canopy. Golding (1982) carried out a series of elegant experiments which demonstrated that, in western Canada at least, the interception process was the dominant mechanism. In the British uplands no similar studies have been carried out; however, this type of experiment may well be inappropriate as the numerous thaw events which characterise the British winter climate would make the interpretation of long-term observations difficult.

To obtain quantitative information on the accumulation and evaporation rates of snow from forest canopies the automated γ-ray system (see Chapter 6) was employed in conjunction with automatic weather stations and two heated plastic-sheet net-rainfall gauges. The system underwent preliminary trials and testing in wet conditions at the Institute of Hydrology's experimental site at Plynlimon, Wales, before being moved to the snowier climate of Aviemore in the Highland Region of Scotland (Figures 8.1 and 8.2).

A 'tree weighing' experiment was also carried out on the same site to measure the rate of snow accumulation and depletion from the canopy of a single tree following the method developed by Roberts (1977) for studying the transpirational responses of trees. Each year a different tree was selected on the basis of its girth being equal to the mean girth of trees growing in the surrounding 100 m^2 area and, unlike the γ-ray experiment, it was possible to take measurements continuously through the winter season.

The analysis of the mass of data that was obtained during these experiments has yet to be completed. The preliminary results and the outline for a snow interception model for forest are presented below.

Figure 8.1 Gamma-ray attenuation studies in the Queens Forest, Aviemore, January 1984. The photograph shows the radioactive source and the meteorological sensors.

8.3 SNOW INTERCEPTION FROM FORESTS—
EXPERIMENTAL RESULTS

The γ-ray attenuation system was installed in 1983 at the Queens Forest site near Aviemore and operated intermittently during the winters of 1983–84 and 1984–85; some 16 storm events, some incomplete, were recorded.

The typical density profiles measured in dry, wet and snow conditions are shown in Figure 8.3. The results, in terms of canopy storage measured by the γ-ray system, and net precipitation

Figure 8.2 Gamma-ray, plastic scintillator detector, supported on the detector tower at the Queens Forest site.

measured by the plastic-sheet gauges, for all the snow events recorded are shown in Figure 8.4.

It is apparent from Figures 8.3 and 8.4 that the water equivalent (WE) of snow accumulated on the canopy can be at least an order of magnitude greater than that recorded in rain conditions and furthermore, that evaporation rates (the difference between the loss rates from the canopy and the net-precipitation rate) can on occasions be large ($= 0.5$ mm h^{-1}); both observations

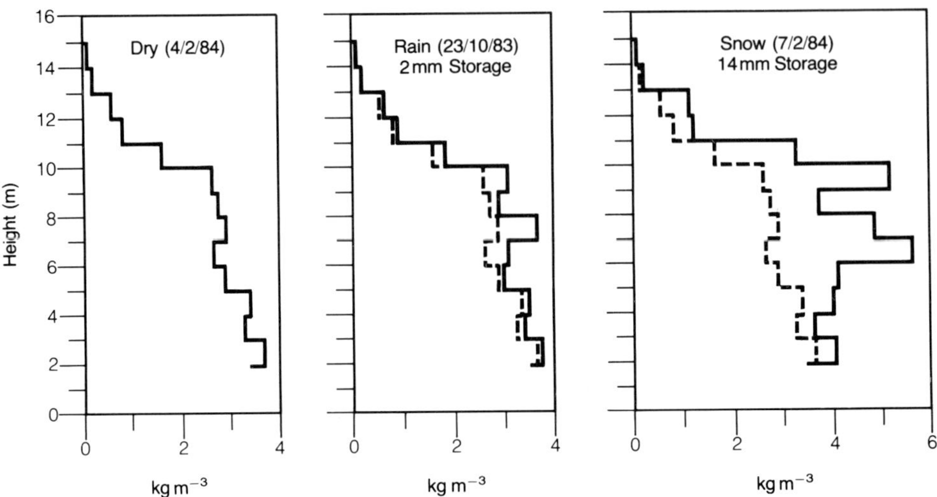

Figure 8.3 Canopy and stem density profiles under dry, wet and snow conditions at the Queens Forest site, Aviemore.

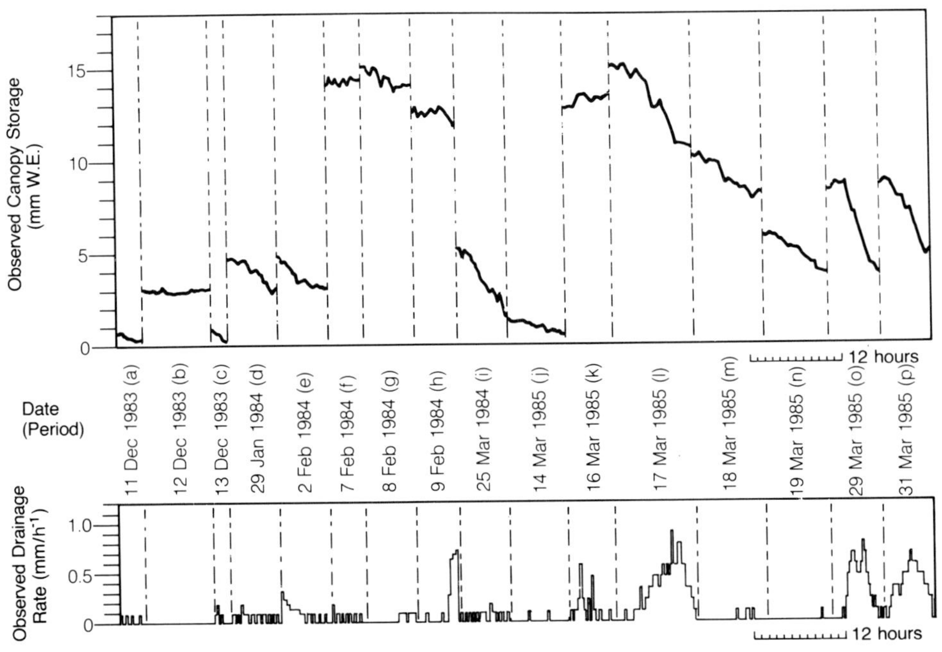

Figure 8.4 Canopy storage and net precipitation measured for all snow events at the Queens Forest, near Aviemore.

demonstrate the potential importance of the snow interception mechanism from forests.

The first snow event recorded, that of 11–13 December 1983, is particularly important and has been the subject of considerable discussion (Calder, 1985b; Morton, 1985). This event demonstrated, conclusively, that high rates of evaporation could occur from precipitation intercepted on forest canopies, even in conditions of low or zero net radiation. It also provided the first results which demonstrated the importance of snow interception as a hydrological process in snowy regions of the UK and gave the basic data for the development of a snow interception model.

8.4 THE AVIEMORE SNOW EVENT OF 11–13 DECEMBER 1983

During the night of 11–12 December 1983 snowfall occurred and a depth of snow equivalent to 3 mm depth of water was measured on the canopy by the γ-ray system (Figure 8.5). The system was operated continuously through the following day, when air temperatures remained close to zero, wind speeds were light (< 2 m s^{-1}), atmospheric specific humidity deficits remained small (0.1 g kg^{-1}), and negligible evaporation rates were recorded.

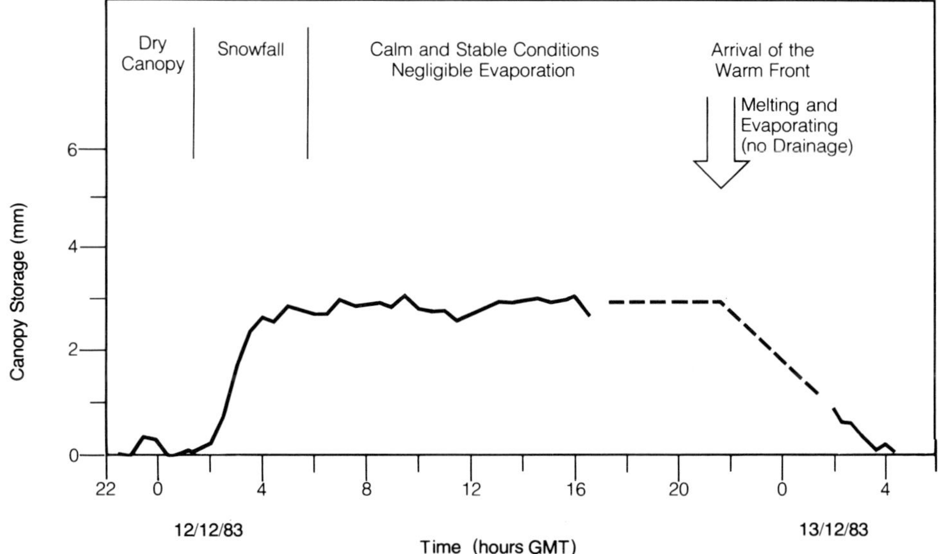

Figure 8.5 Gamma-ray attenuation measurements of the depth of snow (water equivalent) on the forest canopy for the event of 11–13 December 1983. All the snow was lost to evaporation during the night of 12–13 December. (Reproduced by permission from Calder, 1985; copyright Elsevier Science Publishers B.V.)

At 1700 hours the γ-ray system was stopped but the meteorological measurements continued to be logged. At 2100 hours the meteorological conditions began to change; a frontal system bringing in warm maritime air increased the air temperature to 4.5 °C, increased atmospheric humidity, and (because of the increased temperature, see Figure 8.6) increased specific humidity deficits to 1.0 g kg^{-1}. The arrival of the front was associated with high windspeeds of 10–15 ms^{-1}. Net radiation was close to zero throughout the night. The γ-ray measurements were recommenced at 0200 hours on 13 December and evaporation rates of 0.5 mm h^{-1} were recorded by the system. At no time during the night was significant drainage from the canopy recorded, either

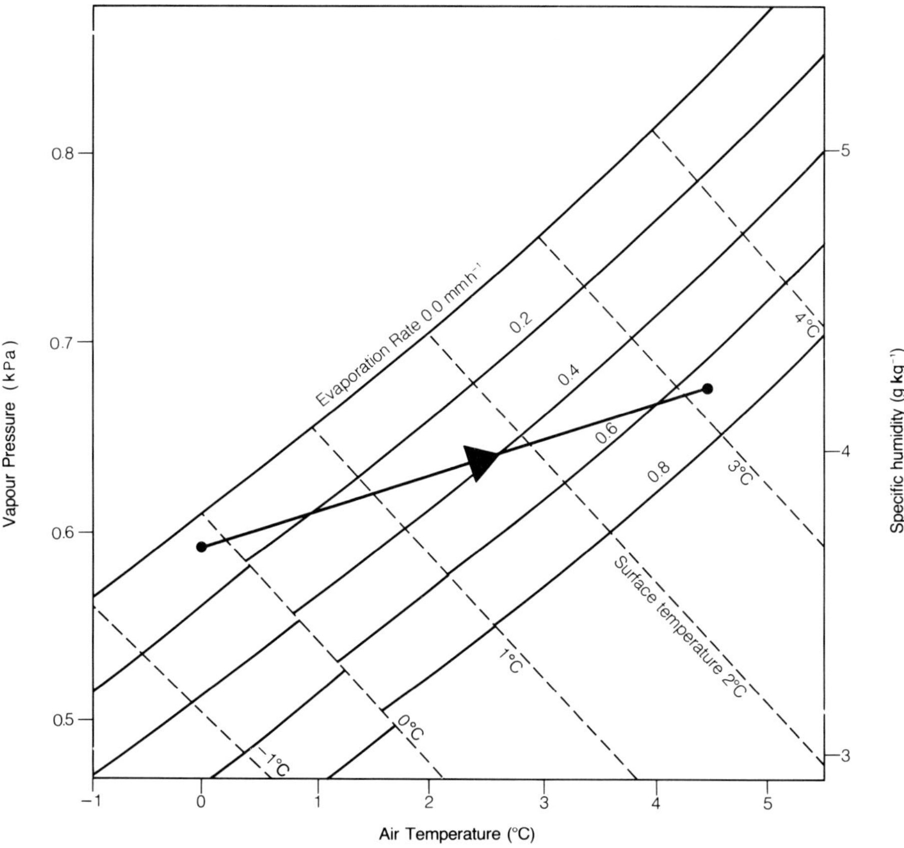

Figure 8.6 Evaporation rates of intercepted precipitation and surface temperatures calculated from solution of the combination equation (equations 2.23a, 2.24a, 2.19a, 2.21a), from forests (r_a ≈ 3.5 s m^{-1}) as a function of atmospheric humidity and air temperature when the net radiation is zero. The changes in meteorological conditions observed during the arrival of the warm front on the night of 12 December 1983 at Aviemore are shown by an arrow. (The effects of surface temperature 'clamping' at the freezing point during snow melt or re-freezing conditions has not been taken into account.)

by the plastic-sheet net-rainfall gauges (<0.15 mm) or by visual observations.

The observations recorded during the different stages of this event were later found to be typical of many of the events recorded at the Aviemore site and served to provide a basis for understanding the mechanisms involved. Essentially these are that during calm, near-freezing atmospheric conditions, when small specific humidity deficits pertain and when the canopy is covered by a 'smooth' snow cover, evaporation rates are very small; so small that it would appear that the aerodynamic resistance is likely to be much higher than the average wet condition value of 3.5 s m^{-1}. In the conditions (possibly brought about by the arrival of a warm front) of higher wind speeds and warmer air, there is sufficient energy and an adequate transport mechanism for melting and evaporating intercepted snow. Figure 8.6 shows that the surface temperature as calculated by solution of the combination equation (equations 2.19a and 2.21a) would have been below freezing prior to the arrival of the warm front, rising to about 3.0 °C subsequently. For parts of the canopy holding snow the surface temperatures would be 'clamped' at the melting point at a specific humidity of 3.8 g kg^{-1}. This value is less than that of the air after the arrival of the warm front (4.2 g kg^{-1}) and consequently *condensation* would have occurred on to the melting snow surface until such time that the particular patch of snow had melted and the surface temperature increased to the point at which the specific humidity of the surface exceeded that of the air and evaporation could take place. It seems likely that during the melting both evaporation and condensation would be taking place concurrently from different patches of snow at different depths within the canopy.

8.5 SNOW INTERCEPTION FROM FORESTS—
MODELLING

8.5.1 Methodology

For many reasons, the detailed modelling of snow accumulation and depletion on a forest canopy is a formidable task: the varying wind speed and temperature during snowfall may affect the quantity of snow retained on the canopy; evaporation may occur as sublimation from the snow or from melt water retained on the leaves; and the distribution of snow and melt water may be very inhomogeneous throughout the canopy. Successive periods of accumulation, sublimation and melt in different sequences will increase this inhomogeneity. Furthermore, it is readily observed that, during the melt phase, large balls of snow may be displaced from the canopy in addition to drips of meltwater.

So far no attempt has been made to investigate the phenomenon to this degree of detail; instead, a broader more pragmatic perspective has been taken in an attempt to understand the physical basis for the major and underlying mechanisms with the objective of producing a practical, physically based model for assessing snow interception losses over whole winter seasons. Snow accumulation on the canopy is treated with a 'build-up function'; the mass and energy transfers and the changes in mass storage on the tree canopy are considered to occur in one of the three states termed sublimation, melt and liquid and are defined by both the calculated temperature of the surface and the depth of solid water storage. The release of liquid water from storage within the snow is then assumed to form an effective input (precipitation minus evaporation) to a conventional rainfall interception model (section 3.2).

8.5.2 Snow build-up function

For the snow events recorded at the Aviemore site a well-defined relationship exists between the water equivalent depths of snow (per unit ground area) lying on both the canopy and the ground following snow precipitation (Figure 8.7). This relationship was used to calibrate the snow storage build-up equation, which relates the rate of snow build-up, dc_S/dt (where the subscript S

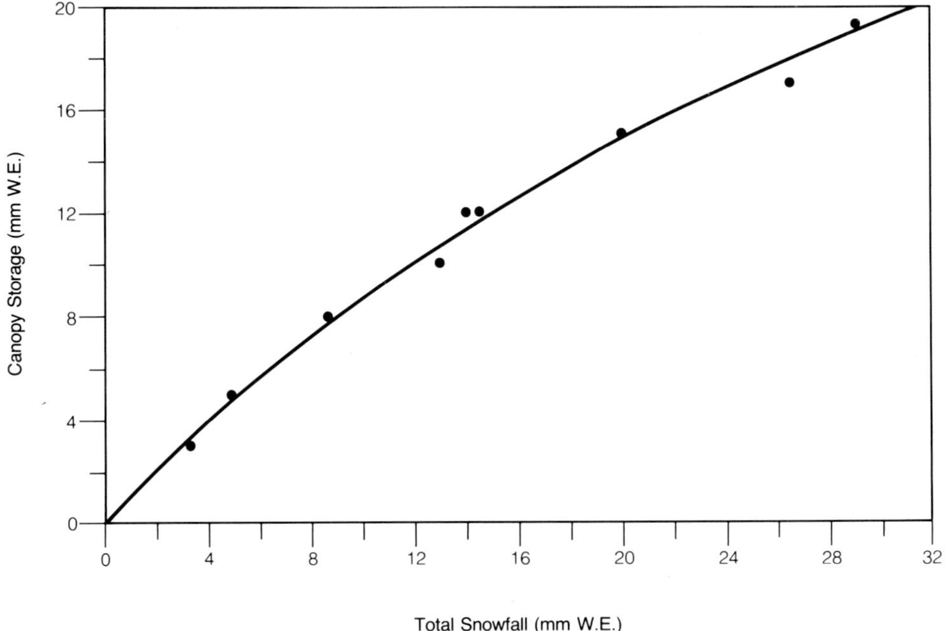

Figure 8.7 The relation between canopy storage and total snowfall, measured immediately after precipitation, which was used to calibrate the 'snow build-up' function.

denotes snow or solid phase conditions), to the rate of precipitation, dP/dt:

$$dc_S/dP = (dc_S/dt)\,(dP/dt) = 1 - c_S/B \qquad 8.1$$

The snow build-up parameter B can be interpreted as the maximum water equivalent depth of snow (per unit ground area) capable of being held on the canopy.

8.5.3 Transfer coefficients

For events when the snow passed through the melt phase leaving a predominantly liquid surface on the canopy, an aerodynamic resistance, r_{aL} (where the subscript L denotes liquid conditions), of 3.5 s m^{-1}, equal to that previously determined in rain events (Calder, 1977), proved adequate when used with the conventional combination equation in describing the measured loss rates.

In sublimation conditions the aerodynamic resistance, r_{aS} (where the subscript S denotes snow or solid phase conditions), was determined by optimization to be 35 s m^{-1}. The higher value in sublimation conditions probably arises from both atmospheric and surface factors; atmospheric conditions are generally slightly more stable and wind speeds less but, perhaps more importantly, the surface of the snow on the tree is much smoother and occupies a much reduced total surface area than in rain conditions, conceivably by as much as the ratio of the aerodynamic resistances in the two conditions.

The treatment of the transition between these two states is the subject of current studies; in reality it will be gradual and will depend not only on the total amount of snow storage on the canopy but also on the way in which it is distributed and on the meteorological conditions. However, for the pragmatic approach adopted here, an abrupt transition is defined in terms of a mean total storage capacity, c^T, and whether calculated estimates of the surface temperature (T_{oL} and T_{oS}, see below) are above or below the freezing point temperature. Thus:

$$r_a = r_{aS} \text{ if } c_S + c_L + c > c^T \text{ or } (T_{oL} < 0 \text{ or } T_{oS} < 0) \qquad 8.2$$

and

$$r_a = r_{aL} \text{ if } (c_S + c_L + c \leqslant c^T) \text{ and } (T_{oL} \geqslant 0 \text{ and } T_{oS} \geqslant 0) \qquad 8.3$$

where:

c_S = water equivalent depth, per unit ground area, of snow
c_L = depth, per unit ground area, of liquid held in the snow
c = depth, per unit ground area, of liquid held on the canopy

8.5.4 Surface temperature

Two estimates of surface temperature were calculated from solution of the combination equation using the assumption that the surface was initially either in the solid or the liquid state. If a liquid state is assumed initially, together with an aerodynamic resistance appropriate to this state, the surface temperature estimate, T_{oL}, is given by equation 2.19a:

$$T_{oL} = T + \frac{Hr_{aL}/\rho\lambda - \delta q}{\Delta' + c_p/\lambda} \qquad 8.4$$

If a solid state is assumed initially, together with an aerodynamic resistance appropriate to this state and a term to take into account the extra latent heat of fusion required for evaporation from the solid phase, the surface temperature estimate, T_{oS}, is given by equation 2.21a:

$$T_{oS} = T + \frac{Hr_{aS}/\rho(L + \lambda) - \delta q}{\Delta' + c_p/(L + \lambda)} \qquad 8.5$$

8.5.5 Canopy storage mass balance in sublimation, melt and liquid conditions

Sublimation condition

Criteria: $T_{oL} < 0\,°C$ or $T_{oS} < 0\,°C$

Evaporation rates, E, from the solid phase can be calculated from the combination equation, modified to take into account the latent heat of melting (equation 2.24a) i.e.:

$$E = \frac{\Delta'H + \rho c_o \delta q/r_{aS}}{(L + \lambda)(\Delta' + c_p/(L + \lambda))} \qquad 8.6$$

No melt will occur in this condition but if liquid water is held within the snow the temperature will be held at the melting point and the liquid will freeze at a rate $-M$, where M is determined by the energy balance equation:

$$H = (L + \lambda)E + C + LM$$

ie.

$$M = (H - (L + \lambda)E - C)/L \qquad 8.7$$

where E is calculated from the definition equation for aerodynamic transfer, i.e.:

$$E = \rho c_p (q_s(0) - q)/r_a \qquad 8.8$$

and sensible heat, C, is given by:

$$C = \rho c_p (0 - T)/r_a \qquad 8.9$$

The finite difference approximation of the change in snow storage on the canopy is then given by:

$$\Delta cs = \frac{dc_S \Delta P}{dP} - M\Delta t - E\Delta t$$

$$= (1 - B/c_S)\Delta P - M\Delta t - E\Delta t \quad \text{if} -M\Delta t < c_L \qquad 8.10$$

$$= (1 - B/c_S)\Delta P + c_L - E\Delta t \quad \text{if} -M\Delta t \geqslant c_L \qquad 8.11$$

where Δt = a finite time increment.

The change in liquid storage within the snow is given by:

$$\Delta c_L = M\Delta t \quad \text{if} -M\Delta t < c_L \qquad 8.12$$

$$\Delta c_L = - c_L \quad \text{if} -M\Delta t \geqslant c \qquad 8.13$$

Melt condition

Criteria: $T_{oL} > 0\,^\circ C$ and $T_{oS} > 0\,^\circ C$ and $c_S > 0$

Evaporation rates from a melting snow surface were again calculated from the definition equation for aerodynamic transfer (equation 8.8). Melt was assumed to occur at a rate M, determined by the energy balance equation (8.7) with E defined by equation 8.8 and sensible heat given by equation 8.9.

The finite difference approximation of the change in snow storage of the canopy is then given by:

$$\Delta c_S = \frac{dc_S \Delta P}{dP} - M\Delta t - E\Delta t$$

$$= (1 - B/c_S)\Delta P - M\Delta t - E\Delta t \ \text{if} \ ((1 - B/c_S)\Delta P - M\Delta t - E\Delta t)$$

$$+ c_S > 0 \qquad 8.14$$

$$= -c_S \qquad \text{if} \ ((1 - B/c_S)\Delta P - M\Delta t - E\Delta t) + c_S \leqslant 0 \quad 8.15$$

The change in liquid storage within the snow is given by:

$$\Delta c_L = M\Delta T \qquad\qquad \text{if } c_L + M\Delta T < p_c c_S \qquad 8.16$$

$$\Delta c_L = p_c c_L - M\Delta t \qquad\qquad \text{if } c_L + M\Delta t \geq p_c c_S \qquad 8.17$$

$$Q_s = (M\Delta T - \Delta c_L)/\Delta t \qquad\qquad 8.18$$

where:

p_c = the maximum ratio of liquid to solid storage, which is generally accepted to be, and is assumed here to be, 0.15, and

Q_s = the effective input rate to a conventional rainfall interception model.

Liquid condition

Criteria: $T_{oL} > 0\ °C$ and $T_{oS} > 0\ °C$ and $c_S = 0$

In this condition the model reverts to a conventional rainfall interception model (see section 6.7).

Figure 8.8 Snow being redistributed from the topmost branches of the canopy during high wind-speed conditions.

8.6 MODEL PREDICTIONS

For the majority of storms recorded the above simplistic approach described the general features of the observations surprisingly well. However, further studies are necessary to investigate these processes in more detail, including the radiative exchange and the effects of snow distribution within the canopy on evaporation and melt rates.

The model adequately describes the high rates of evaporation that are sometimes observed and gives credence to the hypothesis that the difference in snowfall recorded beneath forest canopies and in clearings is principally due to interception rather than redistribution. Direct observations in high wind-speed conditions also support this view; even though conditions have been observed where snow is being redistributed on the canopy (Figure 8.8), no measurable net loss was observed with the γ-ray experiment.

Both the observations and the model confirm that, as in Canadian forests, snow interception is a significant hydrological process and indicate that, for the British uplands, interception in snow conditions may be in proportional terms as large, if not larger, than that in wholly liquid conditions.

CHAPTER 9

Predicting the Effects of Land Use Change— Evaporation Models

9.1 INTRODUCTION

At the simplest level the effects of a change in land use may be inferred, if conditions are identical, from observations of the effects of a similar change that has taken place elsewhere. This is the empirical philosophy employed in the application of results from catchment-scale experiments.

In Great Britain, catchment-scale results are of limited value for assessing the effects of a change of land use on evaporation for two reasons. Firstly, and particularly in steep upland areas where snow may form a significant component of the annual precipitation, the experimental uncertainties in the measurement of the input and output variables, precipitation and flow, may be of a similar magnitude even to the long-term difference between these quantities, i.e. the evaporation. Over the past 20 years improvements in techniques and methods for measuring precipitation and flow have not kept pace with the advances in methods and knowledge gained from process studies which attempt more direct measurement of evaporation. There are no indications to suggest that this trend will not continue in the future. Secondly, there will always be some difference between the conditions where studies are carried out and where they are to be applied. Strictly empirical methods provide no insight into the effects of these different site factors; what is required is an understanding of the processes involved, and a means of applying this understanding through a mathematical model.

Perhaps the most sophisticated and rigorous models so far developed involve the use of the Penman–Monteith equation in conjunction with sub-models which take into account both the feedback effects of surface evaporation on atmospheric demand and the stomatal feedback effects between atmospheric demand

and surface evaporation. This approach gives the greatest insight into the mechanisms involved and is quite general in its applicability, but the availability of the detailed meteorological data that are required and our present poor understanding of the evaporation/atmospheric interaction generally restricts its use to research applications. Until this information is available, alternative, semi-empirical methods need to be employed.

Within their range of applicability, somewhat paradoxically, the semi-empirical methods may actually provide more accurate estimates of long-term evaporation losses than the more theoretical methods. This is because of deficiencies in the instrumentation used for measuring the meteorological variables. This is particularly marked when interception losses are being estimated in wet conditions of low humidity deficits, because evaporation estimates calculated from the Penman–Monteith equation then become extremely sensitive to absolute errors in the measurement of wet bulb depressions (which are used for the calculation of the atmospheric humidity deficit term). Calder (1977) showed that a small offset error of 0.3 °C in the wet bulb depression measurement was sufficient to cause an error of about 20% in the long-term interception estimate, an error which is of similar magnitude to the total range of interception loss which is measured from forests in the wet uplands (Figure 7.4).

The two approaches are therefore complementary: the empirical approach is most useful when meteorological data are limited but it is also of value in providing a cross-check on the results of the more complex theoretical methods. The theoretical methods, being more general in their applicability, are of particular value for application to extreme meteorological conditions and can be used to indicate the range of application of the empirical methods.

The development of the theoretical methods has been outlined in the previous sections; the simplification of these and the development of the semi-empirical methods which require minimal data are discussed here.

9.2 RESEARCH MODELS

The development of the research models for investigating the evaporation from wet heather, grass and forest has been treated extensively in Chapter 5 and Chapter 6. In Chapter 8 the development of a model for assessing losses in snow conditions was outlined. A sub-model for estimating the surface resistance of a spruce forest in dry conditions, for estimating transpiration losses, was given by Calder (1977) and is reproduced in section 1.4. Wallace and co-workers (1982) suggested surface resistance values for heather. Taken together this information provides the basis for developing a research model for investigating the evaporation from upland vegetation in all climatic conditions. The experimental observations discussed in the preceding chapters are summarised in terms of the Penman–Monteith parameters, r_a and r_s, in Figure 9.1.

9.3 SIMPLE ANNUAL EVAPORATION MODEL

The approach that was adopted for the development of evaporation models for practical applications, which is outlined below, involved the calculation of evaporation in terms of its two major components, transpiration and interception. Interception was estimated by relation to the amount and duration of precipitation, whilst transpiration was determined in relation to a reference evaporation estimate.

The Penman potential transpiration estimate (equation 2.14) was chosen as the reference because not only are values readily available but because it also gives reasonable estimates of grass evaporation (for which it was calibrated) and of transpiration from spruce forest. The agreement with spruce forest transpiration is not wholly expected. From initial consideration of the Penman–Monteith equation, with typical forest values for r_a and r_s inserted, it appears that the transpiration will be dominated by the atmospheric humidity deficit term rather than the net radiation

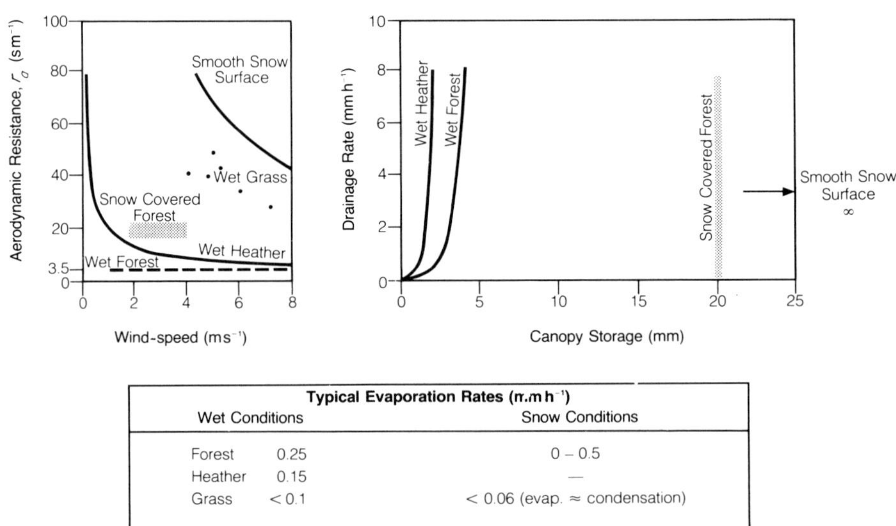

Typical Evaporation Rates (mm h⁻¹)		
Wet Conditions		Snow Conditions
Forest	0.25	0 – 0.5
Heather	0.15	—
Grass	< 0.1	< 0.06 (evap. ≈ condensation)

Figure 9.1 Aerodynamic resistance as a function of wind speed, drainage rates as a function of canopy storage, and typical evaporation rates determined for the different vegetation types in rain and snow conditions. (Reproduced by permission from Calder, 1986; copyright Elsevier Science Publishers B.V.)

term, whereas in the Penman equation the radiative term is usually dominant. However, because spruce exhibits a strong feedback mechanism, whereby with increasing vapour pressure deficit stomata are closed and surface resistance increases (see equation 1.1), the net effect is to effectively 'de-couple' the vapour pressure deficit term so that both the Penman and Penman–Monteith equations give broadly similar values (the Penman E_T estimate usually being about 10% less) when summed over periods of a day or more.

Following this reasoning Calder and Newson (1979, 1980) developed a method, involving a minimal data requirement, for calculating the effects on annual water losses, E_a, brought about by afforesting upland grassland catchments. It was assumed that evaporation losses from grassland and transpiration losses from forests could be estimated from the annual Penman potential transpiration estimate (E_{Ta}), whilst the annual interception loss from the forest could be obtained from a simple function involving annual rainfall (P_a), ie:

$$E_a = E_{Ta} + f(P_a\alpha - w_a E_{Ta})\qquad 9.1$$

where:

α = the interception fraction (35–40% for regions of the UK where annual rainfall exceeds 1000 mm)
w_a = the fraction of the year when the canopy is wet ($\approx$ $0.000122 P_a$)
f = the fraction of the catchment area under canopy coverage.

Use of aerial photographs has shown that, typically, for areas marked on maps as extensive forests, the f value is about 0.66, the remaining area being comprised of roads, rides, riverbanks, clearings and immature plantation with unclosed canopies.

The model indicates that in the wet upland regions of the UK annual evaporation rates from forested catchments (with 75% of their area afforested, equivalent to 50% canopy coverage) may exceed those from grassland by 100% and runoff will be reduced, typically by about 20% (Figure 1.10).

Research on the evaporative characteristics of heather has established that transpiration losses are smaller but interception losses greater than those from grassland. These observations suggest (Calder, 1985a) (Table 9.2) that the annual interception losses from heather can be estimated with an equation of the form:

$$E_a = \beta E_{Ta}(1 - w_a) + \alpha P_a\qquad 9.2$$

where $\beta = 0.5$ and $\alpha = 0.2$.

This equation indicates that similar increases in evaporation and reductions in runoff will be expected when forests replace heather moorland that experiences an annual rainfall of about 1250 mm. For rainfall regimes higher than this the increased interception losses from heather as compared with grass outweigh the reduced transpiration, and the total annual evaporation from heather will be greater than that from grass; the converse is true in lower rainfall regimes.

The approach adopted for the annual model has been developed and extended through the use of a daily interception model to produce a method for estimating evaporation from the major upland vegetation types on a seasonal basis.

9.4 DEVELOPMENT OF THE DAILY INTERCEPTION MODEL

The simplest, one-parameter, model for estimating daily interception loss, I, given only information on daily precipitation, P, is equivalent to that used in the simple annual model, i.e. $I = \alpha P$. Using the mean value of α ($= 0.35$) for all the interception observations discussed in Chapter 7 (see Table 9.1), the percentage root mean square prediction error, $(\Sigma(I^M - I)^2/n)^{1/2} \, (\Sigma I^M/n)$, where I is the predicted and I^M is the measured value for the period and n is the number of periods) for daily, 32-day and 365-day periods are also given in Table 9.1. Clearly, although on an

Table 9.1 Optimised parameters and indices of goodness of fit obtained with the different daily interception models for all sites. The 'RMS error' refers to the root mean square relative prediction error of interception loss for 1-day, 30-day and 365-day periods

Model	γ (or α) (mm)	δ (mm^{-1})	A (mm)	B (mm)	$\bar{E}/\bar{R}$	q	c_{max} (mm)	Sum of squares of deviations (mm^2)	RMS error over period (days)		
									1 (%)	30 (%)	365 (%)
'Constant' 1-parameter	0.35*	–	–	–	–	–	–	6114	86.2	28.5	11.8
Exponential 2-parameter	6.91	0.099	–	–	–	–	–	2967	65.1	24.0	8.7
Logarithmic 2-parameter (Schulze *et al.*, 1977)	–	–	0.27	2.02	–	–	–	2972	65.5	24.1	8.9
Stochastic 3-parameter	–	–	–	–	0.10	1.28	3.62	2907	64.6	23.4	8.9

*This value is the ratio of the total interception loss and total rainfall measured over the whole period of observations. It is not an optimised value.

annual basis the model describes the observations with an error of about 12%, which may be acceptable for most practical purposes, on a daily or seasonal basis this is not likely to be the case.

The fitting of the two-parameter exponential relationship:

$$I = \gamma(1 - \exp(-\delta P)) \qquad 9.3$$

provides a better fit to the observations over all time periods (Table 9.1). The two-parameter, logarithmic model, used by Schulze *et al.* (1977), which takes the form:

$$I = A\log_{10}(P) + B \qquad 9.4$$

where A and B are parameters of the model, also gives a similar goodness of fit on the same data (Table 9.1). (But it will be noted that this mathematical description is unsatisfactory for small P values, which lead to negative predicted interception losses. In the optimisation routine negative values were set to zero.)

The stochastic interception model (see section 3.4) may also be simplified and operated on a daily basis, albeit with an increase in the number of model parameters to three: q, c_{max} and the ratio of $\bar{E}$ and $\bar{R}$. Application of this model, assuming no carry over of the canopy storage from day to day and assuming that evaporation increments, ΔI, are given by the relation:

$$\Delta I = \Delta P \bar{E}/\bar{R} \qquad 9.5$$

results in a small increase in goodness of fit compared with the wholly empirical two-parameter model (Table 9.1). It is arguable whether, for the practical purposes of estimating interception losses in the uplands of the UK, the increased complexity of this model is justified by this rather small improvement in model fit. However, the increased physical realism of the model may be of value in interpreting the effects of a change in management practices which may alter the canopy capacity value (see section 7.3).

Model predictions of daily interception loss at the different sites obtained with the two-parameter daily exponential model, using parameter values optimised over all sites ($\gamma = 6.91$, $\delta = 0.099$), are shown in Figures 9.2, 9.3, 9.4 and 9.5.

9.5 SEASONAL EVAPORATION MODEL FOR FOREST, HEATHER AND GRASS

It is proposed that seasonal estimates of evaporation can be obtained by 'bulking up' daily evaporation estimates, E_d. These

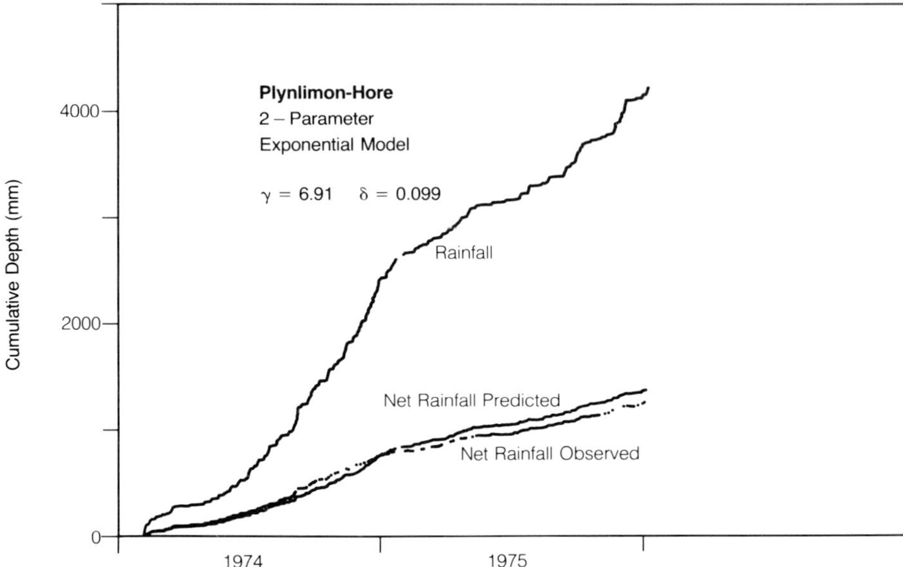

Figure 9.2 Measured cumulative rainfall and interception loss together with predicted interception loss, obtained with the two-parameter daily exponential model using parameter values which were optimised over all sites ($\gamma = 6.91$, $\delta = 0.099$), at the Plynlimon-Hore site.

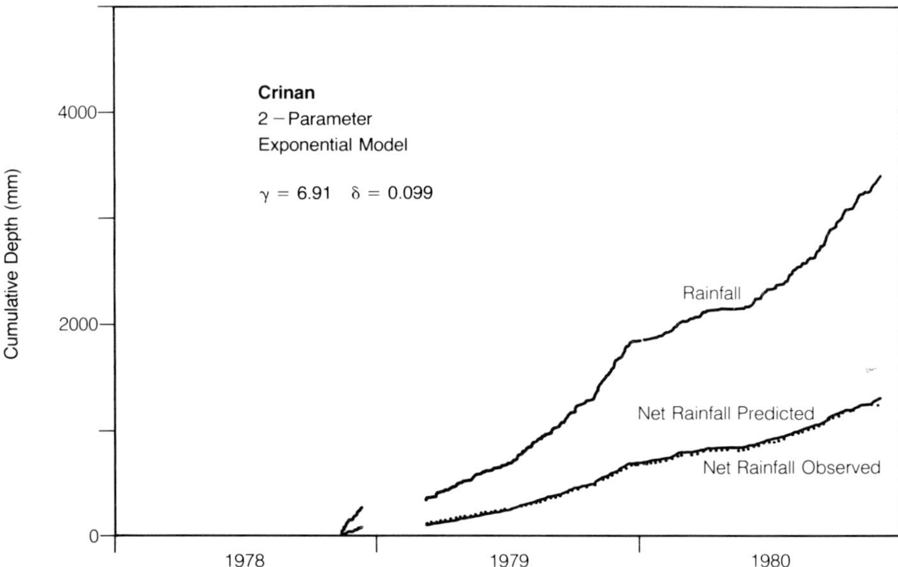

Figure 9.3 Measured cumulative rainfall and interception loss together with predicted interception loss, obtained with the two-parameter daily exponential model using parameter values which were optimised over all sites ($\gamma = 6.91$, $\delta = 0.099$), at the Crinan site.

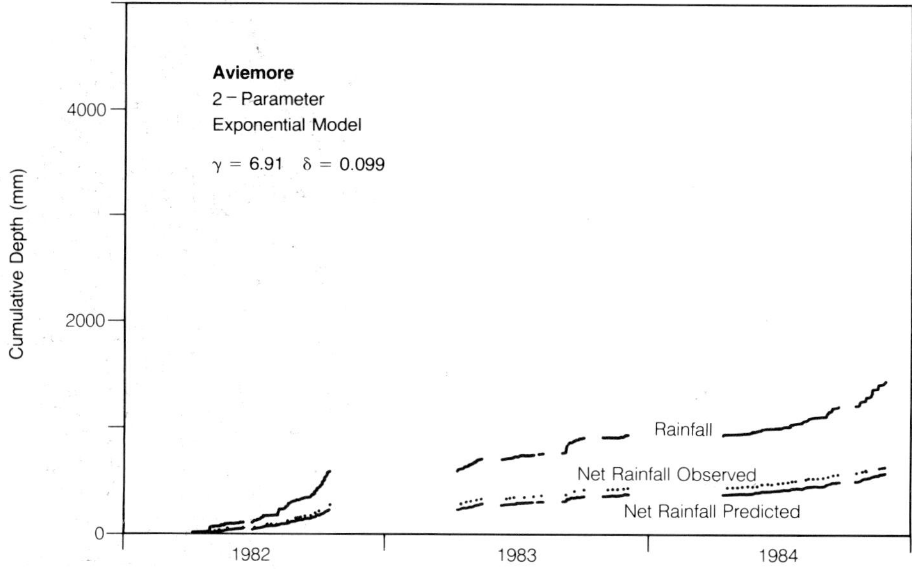

Figure 9.4 Measured cumulative rainfall and interception loss together with predicted interception loss, obtained with the two-parameter daily exponential model using parameter values which were optimised over all sites ($\gamma = 6.91$, $\delta = 0.099$), at the Aviemore site.

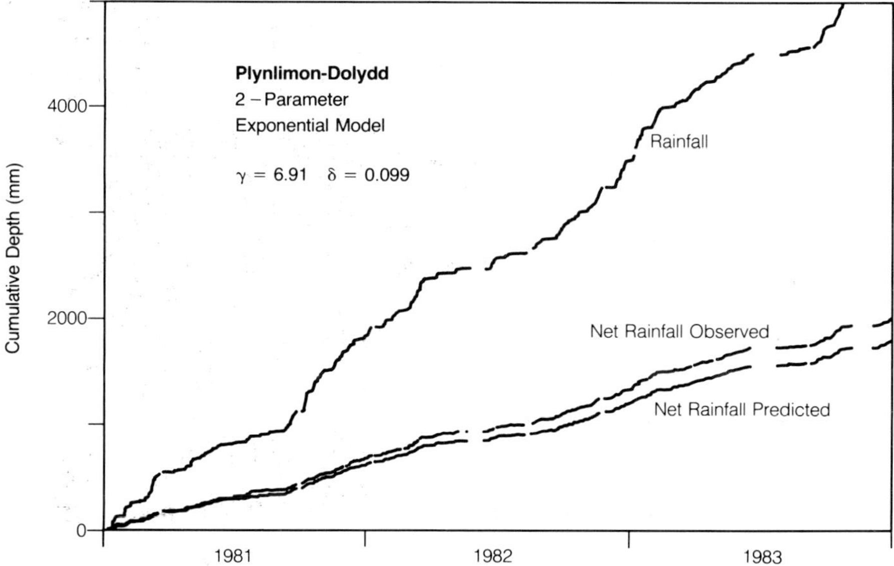

Figure 9.5 Measured cumulative rainfall and interception loss together with predicted interception loss, obtained with the two-parameter daily exponential model using parameter values which were optimised over all sites ($\gamma = 6.91$, $\delta = 0.099$), at the Plynlimon-Dolydd site.

daily values are calculated from the sum of transpiration—which is obtained as the product of the β parameter, a climatologically derived daily Penman E_T estimate and a term $(1 - w)$ which represents the fraction of the day that the canopy is dry—and interception, which is derived from the two-parameter exponential relationship involving daily precipitation, P, i.e.:

$$E_d = \beta E_T (1 - w) + \gamma(1 - \exp(-\delta P))\qquad 9.6$$

where w is the fraction of the day the canopy is wet ($= 0.045P$, for $P < 22$; $= 1$ for $P \geq 22$; after Calder and Newson 1979) and the parameters γ and δ define the form of the interception relationship.

Some support for the use of this form of equation in calculating transpiration is provided by soil moisture deficit modelling studies beneath heather and forest in Scotland (Calder *et al.*, 1984b) and grassland in lowland Britain (Calder *et al.*, 1983b). The interception component has been tested by comparing model predictions of cumulative interception loss, using the mean of all sites value for the γ parameter (from sites in mid-Wales at Dolydd and Hore, the west coast of Scotland at Crinan and the central Highlands of Scotland at Aviemore), with observations at individual sites (Figures 9.2–9.5). The agreement is as good as, if not better than, that expected with the more sophisticated 'research' models using a Rutter-type interception model (Rutter *et al.*, 1971) and à Penman–Monteith evaporation term. The inaccuracy of the research model arises because even with the best local, short-term meteorological measurements currently available (from automatic weather stations), the model sensitivity to errors in the atmospheric humidity deficit term is such that large errors may result; e.g. uncertainties in the measurement of the wet bulb depression of 0.2 °C have been shown to produce errors in the predicted interception loss of 15–20% (Calder, 1977).

Estimates of the α, β, γ and δ parameters for the different vegetation types and the sources from which they were derived are shown in Table 9.2.

The model as outlined is applicable to mature stands of vegetation in climates where snow is not a significant component of the annual precipitation. The treatment of both these contingencies is far from complete; but two working hypotheses have been proposed. For catchments with a high proportion of immature forest Binns (1983, personal communication) suggested that fractional canopy coverage might be represented by an 'S'-shaped function, and such a relationship has been tentatively assumed and used in the Crinan Canal water resource investigation (Chapter 10) with f values as follows: trees aged 0–5 years, 0.1; 6–10 years, 0.33; 11–15 years, 0.75; 16–20 years, 0.95; and 20

Table 9.2 Interception and transpiration observations summarised in terms of the average interception ratio, α, the daily interception model parameters, γ, δ, and the ratio of actual to Penman E_T evaporation, β

| Source | Period | Interception parameters | | | Transpiration fraction (β) |
		α	γ (mm)	δ (mm^{-1})	
Forest					
All sites interception (Plynlimon, Dolydd, Crinan, Aviemore)		0.35	6.9	0.099	
Plynlimon Forest lysimeter	1974–1976	0.30	6.1	0.099	0.9
Dolydd	1981–1983	0.39	7.6	0.099	–
Crinan	1978–1980	0.36	6.6	0.099	–
Aviemore	1982–1984*	0.45	7.1	0.099	–
Heather					
Model estimate derived using automatic weather station data and measured interception parameters (Calder *et al.*, 1986)	1981	–	2.65	0.36	–
Crinan, neutron probe (Calder *et al.*, 1982a, 1984b)	1981–1983	–	–	–	0.58–0.67
Law's heather lysimeters (Calder *et al.*, 1983b)	1964–1968	0.16	–	–	0.47
Sneaton Moor lysimeter (Wallace *et al.*, 1982)	1980	0.19	–	–	0.25–0.5
Grass					
Wye catchment, Plynlimon, indicates total annual evaporation consistent with					1.0

*Not including snow periods.

years or older, 1.0. The results of further studies are awaited to test the validity of this assumption.

At present, and until the detailed analysis of the results from the snow interception study has been completed, it is assumed that the interception losses from forests in both snow and rain conditions are the same proportion of the gross precipitation (the data from which the parameter was derived include some snow periods). The equivalent 'working assumption' for snow interception from heather or grass is that the average daily evaporation rate is zero.

CHAPTER 10

Implications of Upland Afforestation—A Case Study: The Crinan Canal

10.1 INTRODUCTION

The effects on annual runoff of afforesting the catchments which supply the major reservoirs in the UK (assuming previous grass vegetation), have been assessed using the simple annual model (section 9.3) and reported by Calder and Newson (1979). For reservoirs used for hydroelectric power generation, virtually all flows are used for power generation and the total (annual) reduction in runoff is the primary concern. For reservoirs used for supply, for domestic, agricultural or industrial purposes, or for feeding canals, the principal consideration is not with annual totals but whether the catchment-reservoir system is able to meet the demand during the summer months, particularly in a dry year. Here a knowledge of the seasonal differences in the pattern of evaporation is of prime importance, linked with a knowledge of the storage and release regimes of water stored in aquifers on the catchment and the storage and operating rules of the reservoir.

The study of the effects of afforestation on the operation of the Crinan Canal, carried out for the British Waterways Board, provides an example of how the seasonal model may be used in practice and also how it may be used to assess the magnitude of potential future supply difficulties. It also highlights certain areas where our present knowledge of upland evaporation is inadequate, and also, through comparisons with measured canal demand flows and reservoir storage, provides some degree of verification of the model.

10.2 THE CRINAN CANAL RESERVOIRS

Plans for a canal to link Crinan, on the west coast of the Kintyre Peninsula in Scotland, with Loch Fyne were originally drawn up by the eminent Scottish civil engineer, Sir John Rennie. In 1771 James Watt carried out the survey of the route, which would eventually provide a short and sheltered link for shipping which would otherwise have had to make the long and sometimes hazardous trip around the Mull of Kintyre in the journey between the industrial belt of Scotland and the western highlands and islands. It was also intended that the canal would provide both a source of employment in the region and a boost to the local economy. The Crinan Canal Act was passed in 1793 and construction work began the same year under the supervision of Telford; it was completed in 1801.

From the earliest days leaks from some sections of the canal were causing concern to the engineers. In 1816 the canal was greatly improved and in 1823 a section of the canal was re-routed after a bank had failed. Over the years reservoir storage has been increased by raising dam levels and some additional small reservoirs have been installed. This has resulted in the complex pattern of reservoirs, some feeding others, that exists today (Figure 10.1). The canal also intersects and takes water from some natural burns.

Although its original use by commercial shipping and fishing boats has declined, and never at any time has the canal been financially viable in terms of receipts for passage set against operating costs, the usage of the canal and its harbour facilities

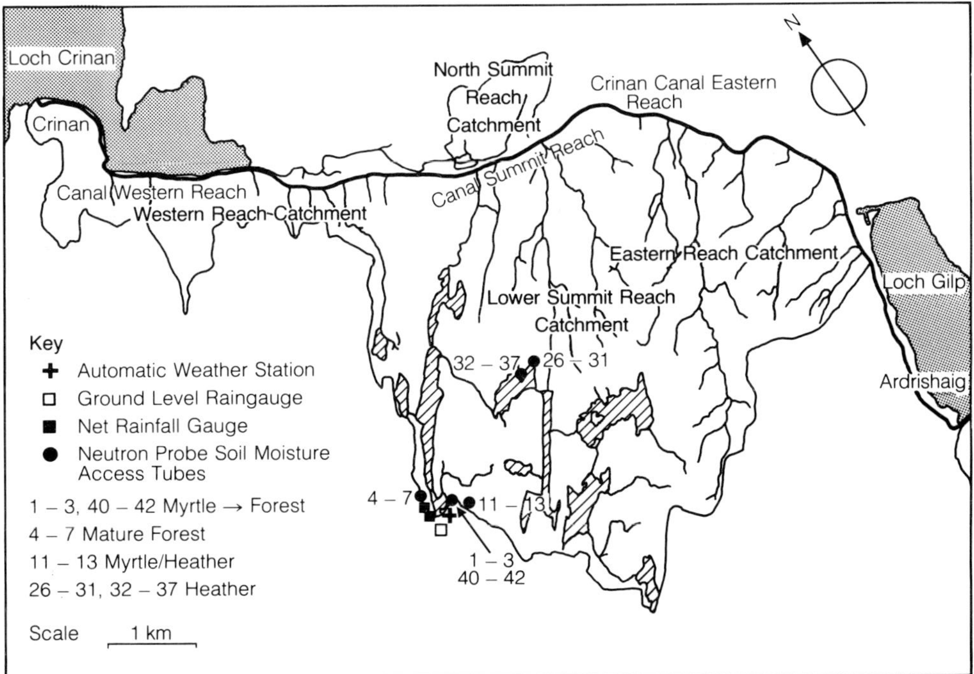

Figure 10.1 Schematic diagram of the Crinan reservoirs.

by pleasure craft is presently increasing and the canal remains vital to the local economy. The safe harbour at Crinan within the sea lock provides a base for the local inshore fishing industry, and the whole canal system provides a major stimulus to tourism in the area.

Despite the efforts to seal leaks in some sections of the canal and to increase the storage of the reservoirs, the water supply to the canal has always been critical during dry summers, and restrictions on its use have sometimes been necessary. Understandably, during the 1970s, when results from evaporation studies were being presented confirming Law's original findings and the Forestry Commission was starting to afforest the land it owned on the catchments of the reservoirs, doubts were raised about the wisdom of this action and concern was expressed about the future viability of the canal. A collaborative project involving the Institute of Hydrology and the British Waterways Board was instigated to assess the future impact of the ongoing afforestation programme.

10.3 EVAPORATION MEASUREMENTS

Three major studies were carried out on the catchment, some of which have been described in earlier chapters.

(1) Plastic-sheet net-rainfall gauge studies of interception loss.
(2) 'Wet-surface' lysimeter studies of the evaporative characteristics of wet heather and grass.
(3) Soil moisture deficit (SMD) measurements on heather moorland (Figure 10.2) and on young and mature forest plantations.

The first two of these experiments were carried out between 1979 and 1982; the third, the soil moisture observations (Figure 10.2), were initiated in 1979 and, as of 1987, are still continuing in the hope they will eventually provide a unique series of measurements of the changing patterns of SMD and evaporation as the young plantation reaches maturity.

10.4 MODELLING STUDIES

10.4.1 Future projections

Two modelling studies were carried out to satisfy different objectives. Firstly, using the annual model (section 9.3) reductions in annual runoff were calculated assuming different forestry rotation periods. Secondly, using the seasonal model (section 9.5)

Figure 10.2 One of the neutron probe access tube sites situated in heather moorland on the Crinan Canal catchments at Kintyre.

and assuming a 'one in ten' drought year, the effects on the supply to the canal on a monthly basis, were calculated for different patterns of afforestation.

10.4.2 Future annual reductions in runoff

The basic information on areas and planting dates for the intended planting schedule (and also for other possible afforestation schedules) for the catchments was provided by the Forestry Commission (Soutar, personal communication) (Table 10.1). Clearly, for some of the short rotation cycles the immature phase, before canopy closure takes place, encompasses almost half the rotation. No interception studies have been reported for the immature phase and, at present, evaporative losses can only be estimated by interpolation. The increase in interception loss, until canopy closure is reached, is represented in the model by an increase in the f parameter. Clearly this is zero for trees of zero age and unity at canopy closure, and the 'S'-shaped function, described in section 9.5, is, in view of the lack of any experimental observations, assumed to represent the intermediate situation. The predicted reductions in runoff, shown in Table 10.1, for the different planting schedules show an increasing effect until a maximum is reached prior to the start of the harvesting of the trees planted in the early 1970s. A cycle of decreasing amplitude

Table 10.1 Predicted increase in evaporation and reduction in runoff as a result of the current and proposed afforestation programme for the Crinan Canal catchments

Year	Annual increase in evaporation from the different age classes expressed as a depth over the whole catchment (mm)					Total annual increase in evaporation (mm)	Reduction in runoff (%)
	0–5	5–10	10–15	15–20	>20		
1960	–	–	–	–	–	–	–
1965	3	–	–	–	–	3	–
1970	–	11	–	–	–	11	1
1975	14	–	25	–	–	40	3
1980	5	47	–	32	–	84	7
1985	10	17	107	–	34	168	13
1990	2	33	39	135	11	221	18
1995	–	8	76	50	153	287	23
2000	–	–	17	96	206	319	26
2005	–	–	–	22	307	329	26
2010	–	–	–	–	330	330	26
2015	1	–	–	–	320	321	26
2020	–	3	–	–	320	323	26
2025	14	–	8	–	177	199	16

would then be expected until the forest reaches a stage of fairly even age distribution. With the planned planting schedule the maximum effect on Canal operations is predicted for the year 2000.

10.4.3 Seasonal effects on runoff

To investigate the seasonal effects on runoff, a simulation was carried out using the actual rainfall measured during 1984 as an input to the model. The year was considered from the point of view of the operation of the canal as a 'one-in-ten' drought year (although the total rainfall for the year was close to the annual mean, the spring and summer periods were particularly dry). Evaporation rates were calculated on a daily basis (using the daily model, section 9.5) for the different proportions of heather moorland and forest cover that were known to exist and that were anticipated for the years 1970, 1985 and 2000. Evaporation from the open water of the reservoir was calculated on the assumption that it was 20% greater than the E_T value (a reverse of the usual procedure of calculating E_T from E_o). Runoff to the reservoirs was assumed to occur from the areas of different vegetation type when rainfall exceeded evaporation and no soil moisture deficit existed; the soil moisture deficit term was accumulated for periods when calculated evaporation exceeded rainfall.

For the purposes of the simulation the reservoirs were considered to act as a single unit whose volume was equal to that of the whole; in effect this is the situation that could only be attained with an ideal operating regime. Using the measured demand curve for lockage and seepage on a weekly basis for

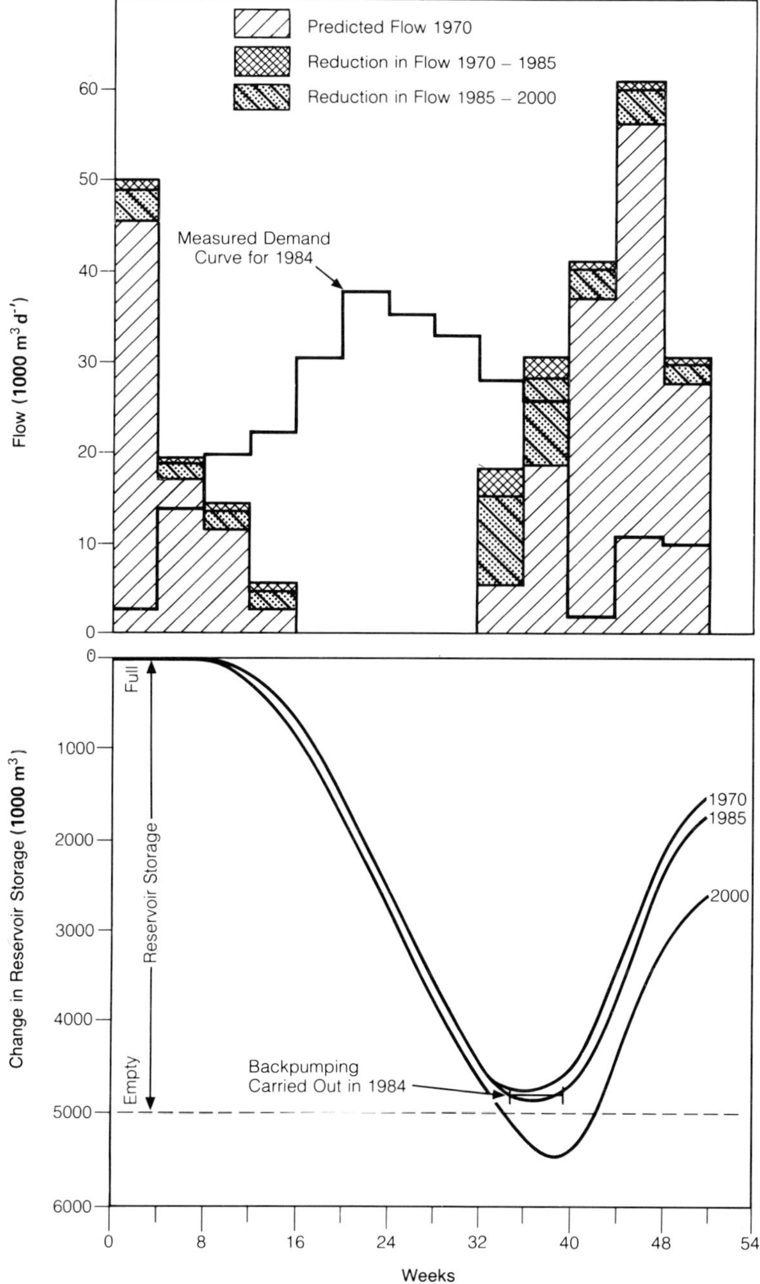

Figure 10.3 Seasonal distribution of runoff and reservoir storage predicted for the Crinan Canal reservoir catchments with different proportions of afforestation, corresponding to the years 1970, 1985 and 2000, given a year with a rainfall distribution equal to that recorded in 1984. The measured demand curve for the reservoir in 1984 is also shown. (Reproduced by permission from Calder, 1985; copyright Elsevier Science Publishers B.V.)

1984 for the summit reach of the canal the drawdown of the summit reservoirs was calculated for the three different afforestation scenarios (Figure 10.3).

Although this treatment of the water supply requirements for the canal is a gross simplification of a complex operational problem, the general features it shows are probably none the less valid. The pattern of afforestation on the catchments in 1970 would not have been sufficient to require additional backpumping, but by the year 2000, with the forestry at its most mature phase, the backpumping requirement will have more than doubled and, perhaps more importantly, it is unlikely that the reservoir would refill following a winter of average rainfall.

CHAPTER 11

Remaining Questions

11.1 INTRODUCTION

Since Law carried out his studies and his provocative results were published, an increasing number of studies have been carried out in the British uplands; in large measure these have vindicated his original findings and have also provided further quantitative data on the evaporation from the different upland vegetation types.

Like the Stocks Reservoir results, these later results were not readily accepted. The results from the Plynlimon forest lysimeter in the wet upland climate of mid-Wales were in some respects more striking than those from Stocks. Evaporation losses (transpiration and interception) totalling about 900 mm, were recorded during the first year of operation, about twice the Penman potential value. They were greeted with disbelief from micrometeorologists more used to evaporation studies in the drier east of the country and, initially, also by Penman.

When similar results were obtained in the succeeding year and a greater understanding was obtained of the mechanisms involved, the results gradually gained acceptance. The integrity of these results depended to a large extent on the accurate estimation of rainfall input above forest canopies in upland areas of steep topography from ground-level raingauge measurements made in a forest clearing, a measurement which cannot easily be checked or verified against other independent methods—so some degree of scepticism was well deserved. The final remnant of doubt that high evaporation rates can occur from wet forests was dispelled, certainly amongst the observers present, when, during the snow event of 11–13 December 1983, about 3.0 mm water equivalent of snow was lost overnight from trees on the Queens Forest γ-ray attenuation site at Aviemore without significant drainage taking place (Figure 8.5). The rate of evaporation achieved overnight and in mid-winter of about 0.5 mm h^{-1}, when a warm frontal system passed over the site, was higher than might reasonably be expected from grass in the lowlands, well supplied with water, on a warm summer's day.

11.2 PRECIPITATION MEASUREMENTS

The precipitation measurement paradox, lucidly explained by Rodda (1971), still remains. There is still no method of measuring to a known degree of accuracy the quantity of precipitation falling at a particular point on the Earth's surface.

The review of methods of precipitation measurement given by Kurtyka (1953) identified six sources of error that raingauges are subject to, namely evaporation, adhesion, colour, inclination, splash and exposure. It was recognised that under the majority of conditions the effects of exposure were by far the most important. Although since then numerous studies, in many countries, have been carried out to develop and compare gauges (see, for example, World Meteorological Organization, 1973) progress has been minimal apart from the efforts made to standardise the use of ground-level gauges to minimise the effects of exposure.

For many practical applications the errors in the measurement of rainfall may not be too serious, but for research applications involving experimental basins and interception studies they are of crucial importance. In the steep upland areas of the UK, where high wind speeds are common, it would be unreasonable to expect the areal rainfall to be measured with an accuracy of better than 5%, even with the use of experimental (high-density) ground-level raingauge networks. In snow conditions the error is likely to be increased at least by a factor of 2 or 3, even if additional measurements involving snow courses, to measure the depth and density of snow, are made.

It can be argued that we are fast approaching the stage, if it has not already been reached, where the long-term precipitation input to an upland catchment can best be obtained from measurements of the runoff and calculation of the evaporation from theoretical considerations which are based on the results of process studies. It will require new, probably revolutionary, methods of precipitation measurement to reverse this trend.

11.3 TRANSPORT THEORY

Although the possibility of genuine site differences within vegetation types makes the detailed comparison of the aerodynamic resistance results obtained using different methods both difficult and confusing, there is, nevertheless, increasing evidence to suggest that a genuine disagreement between methods exists. The results obtained from the γ-ray system, together with results from other, earlier non-micrometeorological studies reported by Calder (1977), Hancock and Crowther (1979), Hancock *et al.*

(1983), Calder *et al.* (1984a) and Wallace *et al.* (1984) are generally mutually consistent within vegetation types; but these results tend to be at variance with estimates of aerodynamic resistance derived from consideration of vegetation height and eddy diffusion theory and (or) with measurements made using micrometeorological methods—eddy correlation (Wallace *et al.*, 1984), Bowen ratio (Miranda *et al.*, 1984).

Differences in both the magnitude and wind-speed dependence of r_a are apparent; the non-micrometeorological methods give values which are smaller and which tend to show less wind-speed dependence than those of the other methods. The reasons for these differences are not yet understood. The non-micrometeorological methods should not be dismissed solely on the grounds of inconsistency; they have the advantage that the calculation of aerodynamic resistance is inherently simpler and requires fewer assumptions than either the theoretical or the micrometeorological methods. Furthermore, the errors associated with the non-micrometeorological methods, although on occasions very large, can usually be identified as such. Real site differences may be partially responsible for the discrepancies: the non-micrometeorological methods have often been used in topographically rough, hilly upland areas, whereas the micrometeorological methods have of necessity only been used in fairly flat locations.

Measurements of the aerodynamic resistance of heather obtained with the 'wet surface' lysimeter from the hilly upland site near Killin (see Chapter 5 and Calder *et al.*, 1984a) suggested an inverse relationship between r_a and wind speed with a power index of 0.6–0.8, as compared with the theoretical value of unity. However, Wallace *et al.* (1984), working in north Yorkshire, and Hall (1985, 1987a), working at sites near Thetford and Aviemore, using lysimetric methods, found that the wind-speed dependence for heather at these relatively flat sites was consistent with the theoretical value of the index. No clear wind-speed dependence has been identified for forests using the γ-ray attenuation system in either wet or snow conditions, or from the earlier modelling studies using observations taken at the Plynlimon forest lysimeter. The lack of dependence may be partially explained if streamlining of the canopy at high wind speed is significant (Fraser, 1962), or if the boundary layer resistance is altered at high wind speeds by a change in flow regime from one in which the flow passes through leaf shoots to one in which the flow is predominantly around the shoot (Grant, 1983, 1984). The existence of another transport mechanism (gust penetration) in addition to eddy diffusion, may also tend to reduce the apparent wind dependence if this other mechanism is not functionally related to wind velocity.

A transport mechanism acting in addition to eddy diffusion would also provide an explanation for the reduced aerodynamic resistance values, obtained with the non-micrometeorological methods (Calder, 1977; Hancock and Crowther, 1979; Calder *et al.*, 1984a; Calder and Wright, 1986), compared with those expected solely from eddy diffusion theory and considerations of vegetation height. Gust penetration in forests has been reported by Denmead (1984) and by Crowther and Hutchings (1984) and a similar mechanism, termed 'clean sweep', has been proposed by Raupach and Legg (1982) as a possible mechanism for explaining their wind tunnel observations of enhanced transfer coefficients for heat and vapour as compared with momentum.

This mechanism, which in hilly upland areas may well be associated with or enhanced by the local topography, might also provide an explanation for the larger discrepancies between observed and theoretically determined values of aerodynamic resistance for heather at the upland Killin site in Scotland as compared with the relatively flat site near Thetford in East Anglia. Clearly further studies are required to investigate this important phenomenon and to develop an adequate theory of its operation.

11.4 SCALE OF THE EVAPORATION–ATMOSPHERE INTERACTION

The areal extent over which high rates of evaporation can be sustained from wet forests is a question at the forefront of present investigations (see section 2.10). The dimensions of UK forests are generally of the order of 5–10 km so it could be argued that the interception studies from UK forests, where interception losses are amongst the highest of those reported anywhere in the world, are only showing an edge effect.

11.5 VARIATION OF EVAPORATION WITH ALTITUDE

11.5.1 Altitudinal effects

Variation in evaporation with altitude is brought about by two principal mechanisms. First, altitudinal variations of the meteorological variables will directly affect meteorological demand. Second, the variation in the meteorological parameters with altitude may indirectly affect transpiration via the influence of the climatic variables on the evaporating characteristics of the crop, both through alterations in live leaf area index and stomatal resistance.

Most evaporation studies in the uplands have been carried out within the altitudinal band where trees would naturally grow (less than about 450 m). Within this range the measurements on grass, forest and heather evaporation can be used with reasonable confidence to predict the change in evaporation resulting from afforestation. Outside this range the evaporation or, more particularly, the transpiration rates from different vegetation types are largely unknown. For completeness of understanding, if not for the more prosaic reason of predicting the effects of afforestation, further studies to investigate evaporation from these higher altitude regions are warranted and have already commenced on the Institute of Hydrology's Balquhidder catchments in Scotland.

11.5.2 Altitudinal variation of the meterological variables

The altitudinal gradients of the meteorological variables, radiation, temperature, wind speed and humidity, which in concert influence atmospheric demand, have not been studied extensively in the British uplands. The measurements that do exist, reported and reviewed by Bleasdale and Chan (1972) and Harding (1978) suggest that the linear gradients that are observed for temperature and humidity in the lowlands may also be extrapolated to the uplands. Harding (1979), using data from automatic weather stations, found a decrease in solar radiation with altitude at a rate of 2.5–3 MJ m^{-2} d^{-1} km^{-1} which was attributed to increased cloud cover associated with regional mean altitude effects (rather than with the altitude of the observing station). There are insufficient direct measurements available to assess altitudinal gradients of net radiation, and estimates based on considerations of the gradients of its components cannot even give, with certainty, the sign of the gradient. As incoming solar radiation decreases with altitude this will tend to make the gradient negative. The albedo, the ratio of the incoming and reflected shortwave radiation, probably does not vary (for the same vegetation) with altitude and will not significantly alter the gradient. However, the net longwave radiation exchange, which is determined by many factors, primarily the surface temperature, the temperature and extent of cloud cover and also the temperature and humidity of the lower atmosphere, probably decreases with altitude and acts to make the gradient of net radiation with altitude positive. Making use of the empirical formula for calculating the longwave exchange used by Penman (1948) and described by Brunt (1930), Harding (personal communication, 1986) calculated the net radiational gradient as -10% km^{-1}. Temperature gradients in Britain are about $-7.2\,°C$ km^{-1}, some variation of the gradient being observed both diurnally and seasonally. Taking an annual average, wind speed increases at a rate of 5.7 m s^{-1} km^{-1}; again the gradients show seasonal and diurnal gradients with maxima in winter and at night. Gradients of humidity show a strong seasonal variation, with a winter minimum value of about

0.5 g kg^{-1} km^{-1} and a maximum of 1.5 g kg^{-1} km^{-1}. Taking into account the temperature and humidity gradients, Harding calculated that specific humidity deficit decreased to about 20% of its sea level value at an altitude of 1000 m although at night and in winter altitudinal gradients above 500 m were essentially zero. Taking all these factors into account the altitudinal gradients of the Penman E_T index of atmospheric demand was found to be about -25% km^{-1}, again with a strong seasonal variation, with the maximum occurring in summer. This figure is broadly in line with that given by the MAFF bulletin on potential transpiration (MAFF, 1967).

Harding's study provides a valuable indication of the long-term gradients that might be expected in the uplands, but little is known of the temporal or spatial variations of these gradients or of the magnitude of the gradients in specific conditions when, for example, vegetation is wet or rainfall is taking place. These are, of course, the conditions when interception losses are occurring and the accurate estimation of the gradients is crucial for the assessment of interception losses. Recent results from the studies carried out by the Institute of Hydrology on the Balquhidder catchments in Scotland (Blackie *et al.*, 1986) also indicate that this issue has not yet been resolved: they found that the gradients of the Penman potential transpiration estimate, instead of decreasing with altitude, actually increased. To a large extent this increase was due to increasing wind speed, but reductions in atmospheric humidity were also sometimes a significant factor.

At an increased level of complexity, the effects of the different evaporative fluxes from the different vegetation types on the atmospheric conditions (particularly in wet conditions) should also be considered but for the foreseeable future both lack of data and lack of knowledge of the operation of the planetary boundary layer preclude this approach for general application.

11.5.3 Altitudinal effects on physiological response

The effects of lowered temperatures on the growth of vegetation are well recognised and documented. It is therefore not unreasonable to expect that there will be commensurate decreases in evaporation with temperature. Two effects have been considered by Scandinavian hydrologists: the limitation that temperature has on growth and the development of leaf area index, and the more direct effect of temperature on stomatal control mechanisms. These have been represented in evaporation models by a regulating function in which transpiration is assumed to be zero when the mean daily temperature is below 5 °C, increases linearly to a maximum value (equal to E_T) when 10 °C is reached and stays at this value until 5 °C is reached in the autumn. Preliminary studies to investigate the application of this type of function to some of the soil moisture measurements discussed in Chapter 4 have shown significantly improved model fits and suggest that further investigations of this type are required to assess fully the evaporation in the high altitude regions of the British uplands.

11.6 SUMMARY

The evaporation rates to be expected from the major upland vegetation types in dry, wet and snow conditions are now fairly well understood. The mechanisms involved, particularly those relating to the feedback effect between evaporation rate and the atmosphere and the areal scale over which this effect takes place, remain largely unknown and await the results of further research.

References

Angstrom, A. (1920). Applications of heat radiation measurements to the problems of the evaporation from lakes and the heat convection at their surfaces. *Geog. Ann.*, **2**, 237–252.

Aston, A. R. (1979). Rainfall interception by eight small trees. *J. Hydrol.*, **42**, 383–396.

Bell, J. P. (1976). *Neutron Probe Practice*. Institute of Hydrology, Report No. 19. Institute of Hydrology, Wallingford, UK.

Binns, W. O. (1980). *Trees and Water*. Arboriculture Leaflet 6, HMSO, London.

Black, T. A., Tanner, C. B. & Gardner, W. R. (1970). Evaporation from a snap bean crop. *Agron. J.*, **62**, 66–69.

Blackie, J. R., Hudson, J. A. & Johnson, R. C. (1986). Upland afforestation and water resources. Preliminary analysis of phase I of the Balquhidder catchment studies. Institute of Hydrology Report. Institute of Hydrology, Wallingford, UK.

Bleasdale, A. and Chan, Y. K. (1972). Orographic influences on the distribution of precipitation. World Meteorological Organisation: Publication No. 326, 322–333.

Brunt, D. (1930). Notes on radiation in the atmosphere. *Quart J. Roy. Meteorol. Soc.*, **58**, 389–420.

Brunt, D. (1939). *Physical and Dynamical Meteorology*. University Press, Cambridge.

Calder, I. R. (1976). The measurement of water losses from a forested area using a "natural" lysimeter. *J. Hydrol.*, **30**, 311–325.

Calder, I. R. (1977). A model of transpiration and interception loss from a spruce forest in Plynlimon, central Wales. *J. Hydrol.*, **33**, 247–265.

Calder, I. R. (1978). Transpiration observations from a spruce forest and comparisons with predictions from an evaporation model. *J. Hydrol.*, **38**, 33–47.

Calder, I. R. (1979a). Do trees use more water than grass? *Water Services*, **83**, 11–14.

Calder, I. R. (1979b). The theory of the neutron probe method. Projeto PNUD/UNESCO/BRA/75/007, Publicacao no. 4., Instituto de Pesquisas Hidraulicas, Porto Alegre, Brazil.

Calder, I. R. (1982). Forest evaporation. *Proc. Can. Hydrol. Symp., 1982*. National Research Council of Canada, Ottawa, pp. 173–193.

Calder, I. R. (1985a). Influence of woodlands on water quantity. *Inst. Biol. Proc. Environ. Div. Symp. on Weather, Woodlands and Water, Edinburgh, 23 March 1984*.

Calder, I. R. (1985b). What are the limits on forest evaporation? Comment. *J. Hydrol.*, **82**, 179–192.

Calder, I. R. (1986). The influence of land use on water yield in upland areas of the U.K. *J. Hydrol.* **88**, 201–212.

Calder, I. R. (1987a). A stochastic model of rainfall interception. *J. Hydrol.*, **89**, 65–72.

Calder, I. R. (1987b). What are the limits on forest evaporation? A further comment. *J. Hydrol.*, **89**, 13–33.

Calder, I. R. & Rosier, P. T. W. (1976). The design of large plastic-sheet net-rainfall gauges. *J. Hydrol.*, **30**, 403–405.

Calder, I. R. & Kidd, C. H. R. (1978). A note on the dynamic calibration of tipping bucket gauges. *J. Hydrol.*, **39**, 383–386.

Calder, I. R. & Newson, M. D. (1979). Land use and upland water resources in Britain—a strategic look. *Water Resources Bulletin* **16**, 1628–1639.

Calder, I. R. & Newson, M. D. (1980). The effects of afforestation on water resources in Scotland. In: Thomas, M. F. & Coppock, J. T. (ed.s) *Land Assessment in Scotland*, Proceedings of the Royal Scottish Geographical Society, Edinburgh, May 1979. Aberdeen University Press, Aberdeen, pp. 51–62.

Calder, I. R. & Neal, C. (1984). Evaporation from saline lakes: a combination equation approach. *Hydrological Sciences Journal*, **29**, 89–97.

Calder, I. R. & Wright, I. R. (1986). Gamma-ray attenuation studies of rainfall interception from Sitka spruce—some evidence for an additional transport mechanism. *Water Resources Research*, **22**, 409–417.

Calder, I. R., Hall, R. L., Harding, R. J., Rosier, P. T. W. & Wright, I. R. (1982a). *Report on Collaborative Project with the British Waterways Board on the Effects of Afforestation on the Runoff from the Catchments Supplying the Crinan Canal Reservoirs.* Institute of Hydrology Report, Institute of Hydrology, Wallingford, UK.

Calder, I. R., Newson, M. D. & Walsh, P. D. (1982b). The application of catchment, lysimeter and hydrometeorological studies of coniferous afforestation in Britain to land-use planning and water management. *Proc. Symp. Hydrolog. Research Basins, Bern 1982*, pp. 853–863.

Calder, I. R., Hall, R. L., Harding, R. J. & Rosier, P. T. W. (1983a). *Upland Afforestation and Water Resources Progress Report, 1982–1983.* Institute of Hydrology Report, Institute of Hydrology, Wallingford, UK.

Calder, I. R., Harding, R. J. & Rosier, P. T. W. (1983b). An objective assessment of soil-moisture deficit models. *J. Hydrol.*, **60**, 329–355.

Calder, I. R., Hall, R. L., Harding, R. J. & Wright, I. R. (1984a). The use of a wet-surface weighing lysimeter system in rainfall interception studies of heather (*Calluna vulgaris*). *J. Clim. and Appl. Meteorol.*, **23**, 461–474.

Calder, I. R., Hall, R. L., Harding, R. J., Wright, I. R. & Rosier, P. T. W. (1984b). *Upland Afforestation and Water Resources Progress Report. Process Studies 1983–1984.* Institute of Hydrology Report, Institute of Hydrology, Wallingford, UK.

Calder, I. R., Hall, R. L., Harding, R. J., Wright, I. R. & Rosier, P. T. W. (1986). *Upland Afforestation and Water Resources Progress Report, 1985–1986.* Institute of Hydrology Report, Institute of Hydrology, Wallingford, UK.

Calder, I. R., Wright, I. R. & Murdiyarso, D. (1987). A study of evaporation from tropical rainforest—West Java. *J. Hydrol.*, **89**, 13–33.

CAS (1980). *Strategy for the U.K. Forest Industry.* CAS Report 6, Centre for Agricultural Strategy, University of Reading, England.

Clayton, C. G. & Cameron, J. F. (1966). A review of the design and application of radioisotope instruments in industry. In: *Radioisotope Instruments in Industry and Geophysics, Vol 1*, International Atomic Energy Agency, Vienna.

Cooper, J. D. (1980). *Measurement of Moisture Fluxes in Unsaturated Soil in Thetford Forest*. Institute of Hydrology Report No. 66, Institute of Hydrology, Wallingford, UK.

Crowther, J. M. & Hutchings, N. J. (1984). Correlated vertical windspeeds in a spruce canopy, In: Hutchinson, B. A., Hicks, B. B., Gay, L. W., Perttu, K. L. & Stewart, J. B. (ed.s) *Forest Environmental Measurement Conference Proceedings*, D. Reidel, Hingham, Mass.

Dalton, J. (1801). On the constitution of mixed gases; on the force of steam or vapour from water and other liquids in different temperatures, both in a Torricellian Vacuum and in air; on evaporation: and on the expansion of gases by heat. *Mem. Lit. Phil. Soc.*, **5**, 366–535.

de Bruin, H. A. R. (1983). A model for the Priestley-Taylor α. *J. Clim. and Appl. Meteorol.*, **22**, 572–578.

Denmead, O. T. (1984). Plant physiological methods for studying evapotranspiration: problems of telling the forest from the trees. *Agric. Water Management*, **8**, 167–189.

Desjardins, R. L., Macpherson, J. L., Alvo, P. & Schuepp, P. H. (1985). In: Hutchinson, B. A., & Hicks, B. B., (ed.s) *The Forest-Atmosphere Interaction*, D. Reidel, Hingham, Mass, pp. 645–658.

Dilley, A. C. (1968). On the computer calculation of vapor pressure and specific humidity gradients from psychometric data, *J. Appl. Meteor.*, **7**, 717–719.

Doorenbros, J. & Pruitt, W. O. (1977). *Crop Water Requirements*. Irrigation and Drainage Paper 24, FAO, Rome, 143 pp.

Ferguson, J. (1952). The rate of evaporation from shallow ponds. *Australian J. Sci. Res.* **5**, 315–330.

Ford, E. D. & Deans, J. D. (1978). The effects of canopy structure, stemflow, throughfall and interception loss in a young Sitka spruce plantation. *J. Appl. Ecol.*, **15**, 905–917.

Fraser, A. I. (1962). Wind tunnel studies of the forces acting on the crowns of small trees, Rep. For. Res., HMSO, London.

Gash, J. H. C. (1979). An analytical model of rainfall interception by forests. *Quart. J. Roy. Meteorol. Soc.*, **105**, 43–55.

Gash, J. H. C. (1985). A study of the fetch requirements of micrometeorological measurements. PhD Thesis, Department of Meteorology, University of Reading, UK.

Gash, J. H. C. & Stewart, J. B. (1977). The evaporation from Thetford Forest during 1975. *J. Hydrol.*, **35**, 385–396.

Gash, J. H. C., Wright, I. R. & Lloyd, C. R. (1980). Comparative estimates of interception loss from three coniferous forests in Great Britain. *J. Hydrol.*, **48**, 89–105.

Goff, S. A. & Gratch, S. (1946). Low-pressure properties of water from -160 to 212 °F. *Trans. Amer. Soc. Heat. Vent. Eng.*, **52**, 95–121.

Golding, D. L. (1982). Snow accumulation patterns in openings and adjacent forest. *Proc. Can. Hydrol. Symp., 1982*. National Research Council of Canada, Ottawa, pp. 91–112.

Grant, R. H. (1983). The scaling of flow in vegetative structures. *Boundary-Layer Meteorol.*, **27**, 171–184.

Grant, R. H. (1984). The mutual interference of spruce canopy structural elements. *Agric. For. Meteorol.*, **32**, 145–156.

Hall, R. L. (1985). Further interception studies of heather using a wet-surface weighing lysimeter system. *J. Hydrol.*, **81**, 193–210.

Hall, R. L. (1987a). The use of a PET microcomputer in rainfall interception studies of heathland. In: Clark, J. A., Gregson, K. & Saffell, R. A., (ed.s) *Computer Applications in Agricultural Environments*, Butterworths, London, pp. 45–56.

Hall, R. L. (1987b). Processes of evaporation from vegetation of the uplands of Scotland, *Trans. Roy. Soc. Edin. (Earth Sciences)*, **78**, 327–334.

Hancock, N. H. & Crowther, J. M. (1979). A technique for the direct measurement of water storage on a forest canopy. *J. Hydrol.*, **41**, 105–122.

Hancock, N. H., Sellers, P. J. & Crowther, J. M. (1983). Evaporation from a partially wet forest canopy. *Ann. Geophys.*, **1**, 139–146.

Harding, R. J. (1978). The variation of the altitudinal gradient of temperature within the British Isles. *Geographisha Annaler*, **60**, 43–49.

Harding, R. J. (1979). Radiation in the British uplands. *J. Appl. Ecol.*, **16**, 161–170.

Harding, R. J. (1986). Exchange of energy and mass associated with a melting snowpack. *Proc. Budapest Symp.*, *IAHS Publ.*, **155**, 3–15.

Harding, R. J. & Rosier, P. T. W. (1984). *Snow Interception*. Institute of Hydrology Internal Report, Institute of Hydrology, Wallingford, UK.

Herwitz, S. R. (1985). Interception storage capacities of tropical rainforest canopy trees. *J. Hydrol.*, **77**, 237–252.

Hubbell, J. H. (1969). *Photon Cross Sections, Attenuation Coefficients and Energy Absorption Coefficients from 10 keV to 100 GeV*. Rep. NSRDS-NBS 29, U.S. Dep. of Commer., Natl. Bur. of Stand., Washington, D.C.

Jackson, I. J. (1975). Relationships between rainfall parameters and interception by tropical forest. *J. Hydrol.*, **24**, 215–238.

Jarvis, P. G. & Stewart, J. B. (1979). Evaporation of water from plantation forest. In: Ford, E. D., Malcolm, D. C. & Atterson, J. (ed.s) *The Ecology of Even Aged Forest Plantations*, Inst. Terrestrial Ecology, NERC, Cambridge, pp. 327–350.

Jarvis, P. G., Edwards, W. R. N. & Talbot, H. (1981). Models of plant and crop water use. In: Rose, D. A. & Charles-Edwards, D. A. (ed.s) *Mathematics and Plant Physiology*, Academic Press, New York, pp. 151–194.

Jeffree, C. E., Johnson, R. P. C. & Jarvis, P. G. (1971). Epicuticular wax in the stomatal antechamber of Sitka spruce and its effects on the diffusion of water vapour and carbon dioxide. *Planta*, **98**, 1–10.

Kurtyka, J. C. (1953). *Precipitation Measurements Study*. Report of Investigation No. 20, State Water Survey, Illinois, 178 pp.

Law, F. (1956). The effect of afforestation upon the yield of water catchment areas. *J. Br. Waterworks Assoc.*, **38**, 489–494.

Law, F. (1957). Measurement of rainfall, interception and evaporation losses in a plantation of Sitka spruce trees. IUGG/IASH, General Assembly of Toronto, **2**, 397–411.

Leyton, L., Reynolds, E. R. C. & Thompson, F. B. (1967). Rainfall interception in forest and moorland. In: Sopper, W. E. & Lull, H. W. (ed.s) *International Symposium on Forest Hydrology*, Pergamon Press, Oxford, pp. 163–168.

Linacre, E. T. (1977). A simple formula for estimating evaporation rates in various climates using temperature data alone. *Agric. Meteorol.*, **18**, 409–424.

List, R. J. (1958). *Smithsonian Meteorological Tables, Sixth Revised Edition*. The Smithsonian Institution, Washington, D.C., 527 pp.

MAFF (1967). *Potential Transpiration*. Technical Bulletin No. 16, Ministry of Agriculture, Fisheries and Food, London.

Massman, W. J. (1980). Water storage on forest foliage: A general model. *Water Resources Research*, **16**, 210–216.

Massman, W. J. (1983). The derivation and validation of a new model for the interception of rainfall by forests. *Agric. Meteorol.*, **28**, 261–286.

McCulloch, J. S. G. (1965). Tables for the rapid computation of the Penman estimate of evaporation. *East African Agriculture and Forestry Journal*, **30**, 286–295.

McNaughton, K. G. (1976). Evaporation and advection II: Evaporation downwind of a boundary separating regions having different surface resistances and available energies. *Quart. J. Roy. Meteorol. Soc.*, **102**, 193–202.

McNaughton, K. G. & Jarvis, P. G. (1983). Predicting the effects of vegetation changes, In: Kozlowski, T. T. Water Deficits and Plant Growth, Vol. VII, Academic Press Inc., pp. 1–47.

McNaughton, K. G. & Spriggs, T. W. (1986). A mixed-layer model for regional evaporation. *Boundary-Layer Meteorol.*, **34**, 243–262.

Miranda, A. C., Jarvis, P. G. & Grace, J. (1984). Transpiration and evaporation from heather moorland. *Boundary-Layer Meteorol.*, **28**, 227–243.

Monteith, J. L. (1965). Evaporation and environment. *Symp. Soc. Exper. Biol.*, **19**, 205–234.

Monteith, J. L. (1973). *Principles of Environmental Physics*. Edward Arnold, London.

Monteith, J. L. (1977). Resistance of a partially wet canopy: whose equation fails? *Boundary-Layer Meteorol.*, **12**, 379–383.

Monteith, J. L. (1981). Evaporation and surface temperature. *Quart. J. Roy. Meteorol. Soc.*, **107**, 1–27.

Morton, F. I. (1984). What are the limits on forest evaporation? *J. Hydrol.*, **74**, 373–398.

Morton, F. I. (1985). What are the limits on forest evaporation? Reply. *J. Hydrol.*, **82**, 184–192.

Murray, F. W. (1967). On the computation of saturation vapour pressure. *J. Appl. Meteorol.*, **6**, 203–204.

Oliver, H. R. (1975). Wind speeds within the trunk space of a pine forest. *Quart. J. Roy. Meteorol. Soc.*, **101**, 167–168.

Oliver, H. R. (1983). The availability of evaporation data in space and time for use in water balance computations. *Proceedings of the Hamburg Workshop, August 1983.* IAHS Publication No. 148.

Olszyczka, B. (1979). Gamma-ray determinations of surface water storage and stem water content for coniferous forests. PhD Thesis, Department of Applied Physics, University of Strathclyde, Scotland.

Penman, H. L. (1948). Natural evaporation from open water, bare soil and grass. *Proc. Roy. Soc. Ser. A*, **193**, 120–145.

Priestley, C. H. B. (1966). The limitation of temperature by evaporation in hot climates. *Agric. Meteorol.* **3**, 241–246.

Priestley, C. H. B. & Taylor, R. J. (1972). On the assessment of surface heat flux and evaporation using large scale parameters. *Mon. Weather Rev.*, **100**, 81–92.

Raupach, M. R. & Legg, B. J. (1982). The use and limitations of flux-gradient relationships in meteorology. *Agric. Water Management*, **8**, 119–131.

Rijtema, P. E. (1965). *An Analysis of Actual Evapotranspiration*. Agricultural Rep. No. 659, Centre for Agricultural Publications and

Documentation, Wageningen, The Netherlands, 107 pp.

Roberts, J. M. (1977). The use of "tree cutting" techniques in the study of the water relations of mature *Pinus sylvestris* L.: 1. The technique and survey of the results. *J. Exper. Bot.*, **28**, 751–767.

Roberts, J. M. (1978). The use of the "tree cutting" technique in the study of the water relations of Norway spruce (*Picea abies* (L.) Karst.). *J. Exper. Bot.*, **29**, 465–471.

Roberts, J. M. (1983). Forest transpiration: a conservative hydrological process? *J. Hydrol.*, **66**, 133–141.

Rodda, J. C. (1971). *The Precipitation Measurement Paradox—the Instrument Accuracy Problem*. WMO/IHD Report No. 16, World Meteorological Organization.

Rutter, A. J. (1963). Studies in the water relations of *Pinus sylvestris* in plantation conditions. I. Measurements of rainfall and interception. *J. Ecol.*, **51**, 191–203.

Rutter, A. J. (1966). Studies on the water relations of *Pinus sylvestris* in plantation conditions. *J. Appl. Ecol.*, **3**, 393–405.

Rutter, A. J., Kershaw, K. A., Robins, P. C. & Morton, A. J. (1971). A predictive model of rainfall interception in forests. I: Derivation of the model from observations in a plantation of Corsican pine. *Agric. Meteorol.*, **9**, 367–384.

Rutter, A. J., Morton, A. J. & Robins, P. C. (1975). A predictive model of rainfall interception in forests. II. Generalization of the model and comparison with observations in some coniferous and hardwood stands. *J. Appl. Ecol.*, **12**, 367–380.

Schulze, R. E., Scott-Shaw, C. R. & Nanni, U. W. (1977). Interception by *Pinus patula* in relation to rainfall parameters. *J. Hydrol.*, **36**, 393–396.

Sellers, P. J. & Lockwood, J. G. (1981). A computer simulation of the effects of differing crop types on the water balance of small catchments over long periods. *Quart. J. Roy. Meteorol. Soc.*, **107**, 395–414.

Shuttleworth, W. J. (1976). A one-dimensional theoretical description of the vegetation–atmosphere interaction, *Boundary-Layer Meteorol.*, **10**, 273–302.

Shuttleworth, W. J. (1977). Comments on "Resistance of a partially wet canopy: Whose equation fails?" *Boundary-Layer Meteorol.*, **12**, 385–386.

Shuttleworth, W. J. & Calder, I. R. (1979). Has the Priestley–Taylor equation any relevance to forest evaporation? *J. Appl. Meteorol.*, **18**, 639–646.

Shuttleworth, W. J., Gash, J. H. C., Lloyd, C. R., Moore, C. J., Roberts, J. M., *et al.* (1984). Eddy correlation measurements of energy partition for Amazonian forest. *Quart. J. Roy. Meteorol. Soc.*, **110**, 1143–1162.

Stewart, J. B. (1977). Evaporation from the wet canopy of a pine forest. *Water Resources Research*, **13**, 915–921.

Stewart, J. B. (1981). Evaporation from forests. *Proceedings of the XVIII IUFRO World Congress, Div. 1, September 1981, Kyoto, Japan*, pp. 130–138.

Stewart, J. B. (1983). A discussion of the relationships between the principal forms of the combination equation for estimating crop evaporation. *Agric. Meteorol.*, **30**, 111–127.

Stewart, J. B. & Thom, A. S. (1973). Energy budgets in a pine forest. *Quart. J. Roy. Meteorol. Soc.*, **99**, 154–170.

Strangeways, I. C. (1972). Automatic weather stations for network operation. *Weather*, **27**, 403–408.

Tan, C. S. and Black, T. A. (1976). Factors affecting the canopy resistance of a Douglas fir forest. *Boundary-Layer Meteorol.*, **10**, 475–488.

Tetens, O. (1930). Uber einige meteorologische Begriffe. *Z. Geophys.*, **6**, 297–309.

Thom, A. S. (1972). Momentum, mass and heat exchange of vegetation. *Quart. J. Roy. Meteorol. Soc.*, **98**, 124–134.

Thom, A. S. (1975). Momentum, mass and heat exchange of plant communities. In: Monteith, J. L., (ed.) *Vegetation and the Atmosphere*, *Vol. 1*, Academic Press, London, pp. 57–109.

Thom, A. S. & Oliver, H.R. (1977). On Penman's equation for estimating regional evaporation. *Quart. J. Roy. Meteorol. Soc.*, **103**, 345–357.

Wallace, J. S., Lloyd, C. R., Roberts, J. M. & Shuttleworth, W. J. (1984). A comparison of methods for estimating aerodynamic resistance of heather (*Calluna vulgaris*) (L) Hull) in the field. *Agric. Meteorol.*, **32**, 289–305.

Wallace, J. S., Roberts, J. M. & Roberts, A. M. (1982). Evaporation from heather moorland in North Yorkshire, England. *Proc. Symp. Hydrolog. Research Basins, Bern 1982*, pp. 397–405.

Walsh, P. D. (1977). The practical analysis and operation of multi-source water supply systems with particular reference to the Lancashire conjunctive use scheme. PhD Thesis, University of Birmingham, UK.

Walsh, P. D. (1980). The impacts of catchment afforestation on water supply interests. *Aqua*, **4**, 82–85.

Walsh, P. D. & Walker, S. (1985). Seasonal variation in water losses from Law's forest lysimeter at Stocks Reservoir. Proceedings of the WRC/WMO Conference on "Effects of land use on fresh waters", University of Stirling, 4–7 June, 1985. In: Solbe, J. F. de L. G. (ed.) *Effects of Land Use on Fresh Waters*, Ellis Horwood, Chichester, UK, pp. 541–545.

World Meteorological Organization (1973). *Annotated Bibliography on Precipitation Measurement Instruments*. WMO/IHD Report No. 17.

Appendix

The calculation of evaporation from solution of the combination equation requires values of the physical properties of air and water vapour. The specific heat of air at constant pressure, c_p, can be treated as a constant within the range -5 to $45\,°C$ with the value of $1010\ \mathrm{J\ kg^{-1}\,°C^{-1}}$. (For practical purposes it is not necessary to take into account the heat content of water vapour in the solution of the combination equation but in rigorous treatments this should be included; the specific heat of water vapour is given as $1890\ \mathrm{J\ kg^{-1}\,°C^{-1}}$.) Physical properties which change by less than 1% per $°C$, such as density, latent heat of vaporisation and the psychrometric constant, can be conveniently estimated from standard tables (Table A.1) (see also Monteith, 1973). Saturated vapour pressure, saturated specific humidity and the rate of change of these quantities with temperature (Δ and Δ') are properties which change more rapidly with temperature. They can also be interpolated from tables, but for computer computation it is more convenient to make use of the equation given by Dilley (1968), which he obtained by drawing on the earlier work of Goff and Gratch (1946), List (1958), Tetens (1930) and Murray (1967).

Using the present symbolism the saturation vapour pressure (kPa) over a water surface can be expressed as:

$$e_s(T) = 0.61078\exp[17.269T/(T + 237.30)]$$

Over an ice surface the saturation vapour pressure is given by:

$$e_s(T) = 0.61078\exp[21.875T/(T + 265.5)]$$

Differentiating the above expressions with respect to temperature yields the gradient relationship above a water surface:

$$\Delta = 2502.9\exp[17.269T/(T + 237.30)]/(T + 237.30)^2$$

and, over an ice surface:

$$\Delta = 3547.3\exp[21.875T/(T + 265.50)]/(T + 265.50)^2$$

The equivalent expressions in terms of specific humidity are:

$$q_s(T) = \epsilon\, e_s(T)/(p - (1 - \epsilon)e_s(T))$$

$$= 0.622e_s(T)/(p - 0.378e_s(T))$$

and

$$\Delta' = \epsilon\Delta p/(p - (1 - \epsilon)e_s(T))^2$$

$$= 0.622\Delta p/(p - 0.378e_s(T))^2$$

Table A.1 Physical properties of air and water vapour (changing by less than 1% per °C)

Temperature (T)		Density of dry air (ρ_a) (kg m^{-3})	Density of air saturated with water vapour at temperature T (ρ) (kg m^{-3})	Latent heat of vaporisation (λ) (J kg^{-1})	Psychrometric constant at an atmospheric pressure of 100 kPa (γ) (kPa °C^{-1})
°C	K				
−5	268.2	1.316	1.314	2,513,000	0.0643
0	273.2	1.292	1.289	2,501,000	0.0646
5	278.2	1.269	1.265	2,489,000	0.0649
10	283.2	1.246	1.240	2,477,000	0.0652
15	288.2	1.225	1.217	2,465,000	0.0655
20	293.2	1.204	1.194	2,454,000	0.0658
25	298.2	1.183	1.169	2,442,000	0.0662
30	303.2	1.164	1.145	2,430,000	0.0665
35	308.2	1.146	1.121	2,418,000	0.0668
40	313.2	1.128	1.096	2,406,000	0.0671
45	318.2	1.110	1.068	2,394,000	0.0675

Subject Index

Aerodynamic resistance
  definition, 17
  of forest, 87
  of heather, 67
  of smooth snow, 100
  of snow-covered forest, 109
  transport theory and, 133
  windspeed dependence for heather
    and grass, 75
Altitudinal effects on evaporation, 56,
  134

Balquhidder experiments
  catchment experiment, 136
  soil moisture observations, 57

Calder and Newson annual upland
    evaporation model, 14
Canopy density measurements, 82
Canopy drainage measurements, 69, 84
Catchment experiments, 5, 136
Combination equation, 19
Crinan canal experiments
  case study, 124
  soil moisture observations, 57

Dalton equation, 17
'Del' approximation, 19
Drop size, effect on interception, 46

Eddy diffusion theory, 29, 133
Energy balance equation, 19
Evaporation
  altitudinal gradient, 134
  evaporation–atmosphere interaction,
    32, 134
  of intercepted rainfall precipitation,
    28, 40, 41, 47, 60, 61, 76, 89
  models, 114–123
  from snow covered vegetation,
    100–113

Forest interception, 89–99, 101, 123
Forest lysimeter, 7
Forest transpiration, 3, 9, 12, 57, 119
Forest (spruce) wetting characteristics,
    72

Gamma-ray attenuation studies
  calibration, 79
  canopy density measurements, 82
  canopy drainage, 84
  dependence of drainage parameters
    and aerodynamic resistance on
    wind speed, 87
  design, 76
  evaporation from partially wet
    canopy, 86
  method, 78
  sites, 81
  snow interception, 102, 105
  wet canopy, 83
Grass drainage lysimeters, 51, 54
Grass interception, 51, 73–75, 123
Grass transpiration, 51, 119, 123
Grass wetting characteristics, 72
Gust penetration, 32

Heather interception, 51, 58, 70–73,
    119
Heather lysimeters, 49, 58
Heather transpiration, 51, 58, 119
Heather wetting characteristics, 72

Interception
  from forests
    gamma-ray attenuation studies, 76
    plastic-sheet net-rainfall gauge
      measurements, 89
  the process, 28, 40, 61
  Rutter model, 40
  from short vegetation (the 'wet
    surface' lysimeter), 60
  simplified 'daily' stochastic model, 47

snow interception, 100
stochastic model, 41

Killin heather interception studies, 66

Land use change, effects of, 1, 3,
        13–15, 114, 124–130
Law's hydrological experiments, Stocks
        Reservoir, 1, 49–54

Models
    Calder and Newson annual (forest
        and grass) evaporation model,
        14
    daily interception models, 118
    Morton 'closed box' model, 37
    Penman equation, 21
    Rutter, 40, 85
    seasonal evaporation model for
        forest, heather and grass, 119
    simple annual evaporation model
        (forest, heather, grass),
        116–118
    simplified 'daily' stochastic, 47
    stochastic, 41–46
    threshold, 85
    two-parameter daily exponential, 94,
        119
Morton 'closed box' model, 37

Neutron probe observations, 9, 58, 123,
        126

Open water evaporation, 22, 128

Penman–Monteith equation, 25
Plastic-sheet net-rainfall gauges
    design, 89
    experimental errors, 93
    sites, 92
    tree thinning experiment, 97
Plynlimon
    catchment experiment, 5
    forest lysimeter, 7
    grass drainage lysimeters, 54
    plastic-sheet net-rainfall
        measurements, 92
    soil moisture observation, 57
    thinning experiment, 97
Priestly–Taylor equation, 24

Queens Forest (Aviemore)
        experiments, 33, 82, 105

Reservoirs, effects of afforestation of
        catchments, 14
Runoff reduction, 3, 127
Rutter model, 40

Smoke bomb studies, 32
Sneaton Moor heather studies, 58
Snow build-up function, 108
Snow interception, 102, 105, 123
Soil moisture observations, 9, 11, 57,
        126
Stochastic model, 41, 47
Stocks Reservoir
    heather and grass lysimeters, 49
    Law's hydrological experiments, 1
Stomatal resistance, 11, 25
Surface resistance sub-model for
        spruce, 11
Surface temperature
    frozen surface, 27, 110
    transpiring surface, 27
    wet surface, 26, 67, 110

Thetford micrometeorological studies,
        11
Thetford soil physical studies, 11
Thinning experiment, 97
Thom and Oliver equation, 23
Tipping bucket flowmeter, 91
Transpiration
    forest, 9, 12, 57
    grass, 51, 54
    heather, 50, 56, 58
Transport theory, 29, 132

Upland afforestation consortium, 15
Upland reservoirs, 13, 125
Uplands, topographic effects, 38, 133

Vapour pressure, 17, 145
Vapour pressure deficit, 20, 145

'Wet surface' lysimeter experiment
    aerodynamic resistance results, 67
    canopy drainage measurements, 69
    evaporation in partially wet
        conditions, 72
    experimental design, 60
    sites, 61
Wetting characteristics of vegetation, 72